Dick
Shonock

CONCENTRATION FUNCTIONS

Probability and Mathematical Statistics

A Series of Monographs and Textbooks

Editors **Z. W. Birnbaum** — *University of Washington, Seattle, Washington*

E. Lukacs — *Bowling Green State University, Bowling Green, Ohio*

---

1. Thomas Ferguson. Mathematical Statistics: A Decision Theoretic Approach. 1967
2. Howard Tucker. A Graduate Course in Probability. 1967
3. K. R. Parthasarathy. Probability Measures on Metric Spaces. 1967
4. P. Révész. The Laws of Large Numbers. 1968
5. H. P. McKean, Jr. Stochastic Integrals. 1969
6. B. V. Gnedenko, Yu. K. Belyayev, and A. D. Solovyev. Mathematical Methods of Reliability Theory. 1969
7. Demetrios A. Kappos. Probability Algebras and Stochastic Spaces. 1969
8. Ivan N. Pesin. Classical and Modern Integration Theories. 1970
9. S. Vajda. Probabilistic Programming. 1972
10. Sheldon M. Ross. Introduction to Probability Models. 1972
11. Robert B. Ash. Real Analysis and Probability. 1972
12. V. V. Fedorov. Theory of Optimal Experiments. 1972
13. K. V. Mardia. Statistics of Directional Data. 1972
14. H. Dym and H. P. McKean. Fourier Series and Integrals. 1972
15. Tatsuo Kawata. Fourier Analysis in Probability Theory. 1972
16. Fritz Oberhettinger. Fourier Transforms of Distributions and Their Inverses: A Collection of Tables. 1973
17. Paul Erdös and Joel Spencer. Probabilistic Methods in Combinatorics. 1973
18. K. Sarkadi and I. Vincze. Mathematical Methods of Statistical Quality Control. 1973
19. Michael R. Anderberg. Cluster Analysis for Applications. 1973
20. W. Hengartner and R. Theodorescu. Concentration Functions. 1973

In Preparation

L. E. Maistrov. Probability Theory: A Historical Sketch

William F. Stout. Almost Sure Convergence

L. H. Koopmans. The Spectral Analysis of Time Series

Kai Lai Chung. A Course in Probability Theory, Second Edition

CONCENTRATION FUNCTIONS

W. Hengartner
R. Theodorescu

Department of Mathematics
Laval University
Quebec, Canada

ACADEMIC PRESS *New York · London · 1973*
A Subsidiary of Harcourt Brace Jovanovich, Publishers

ACADEMIC PRESS, INC.
111 Fifth Avenue, New York, New York 10003

United Kingdom Edition published by
ACADEMIC PRESS, INC. (LONDON) LTD.
24/28 Oval Road, London NW1

Library of Congress Cataloging in Publication Data

Hengartner, Walter.
Concentration functions.

(Probability and mathematical statistics)
Bibliography: p.
1. Concentration functions. 2. Measure theory.
3. Probabilities. I. Theodorescu, Radu, joint author.
II. Title.
QA273.43.H46 519.2'6 73-5297
ISBN 0-12-341050-9

AMS (MOS) 1970 Subject Classifications: 60-02; 60B10; 60F05

PRINTED IN THE UNITED STATES OF AMERICA

Contents

Chapter 3 Convergence Problems

Chapter 4 Concentrations

Chapter 5 Generalizations

Preface

The notion of concentration function was introduced in probability theory in 1937 by Lévy in his famous monograph devoted to sums of random variables. Since then, concentration functions have been used to investigate convergence problems for sums of independent random variables. The material devoted to these functions is to be found scattered throughout the literature. We have not attempted here to write a comprehensive book on this subject. Instead, we have made a selection of basic material in accordance with our own preferences. However, references are given to results that are only mentioned or scarcely touched upon here. In writing this text, we had two basic objectives: to introduce concentration functions to readers with a good background in modern probability theory and to illustrate some applications of these functions, mainly to convergence problems.

The monograph is divided into five chapters. In Chapter 1, a detailed description of the Lévy concentration function and of its basic properties is given. Chapter 2 is devoted to concentration functions of convolutions of probability measures. In Chapter 3, the concentration function is used to investigate convergence problems.

Chapter 4 is devoted to various notions that occur in the literature under the name of concentration, and which are related to concentration functions. Chapter 5 is concerned with generalizations of the concept of concentration function.

The reader may expect to gain four things from this monograph: (1) an overall picture of results concerning concentration functions and related concepts, (2) experience in associating concentration functions with convergence problems, (3) skill in using these functions to interpret various results in probability theory, and (4) some insight into problems concerning sums of random variables.

This monograph, based on notes prepared in 1969 and then successively revised, has been designed in such a manner that it should prove useful for several purposes, namely: (1) as a text for a one- or two-semester graduate course devoted to selected topics in probability theory, (2) as a complementary text to a senior undergraduate course in probability theory, and (3) as a reference book for research workers in probability theory.

Definitions, theorems, lemmas, and propositions are numbered *a.b.c.*, where *a* indicates the chapter and *b* the section. The end of a proof is marked by ◇.

We have made a sincere effort to provide every section with appropriate bibliographical notes and comments; any inaccuracy or omission in assigning priorities is wholly unintentional and deeply regretted. Where it was possible, we inserted additional complements which supplement the theory developed.

This monograph could not have been undertaken without the support provided by the National Research Council of Canada, grants A-7339 and A-7223. We take this occasion to express our deep appreciation for this support. The final draft was prepared during the summer of 1972, when we attended the Summer Research Institute organized by the Canadian Mathematical Congress at Carleton University, Ottawa.

Parts of this monograph were previously distributed, and we are indebted to many friends and colleagues for their valuable advice and criticism. We particularly wish to thank Professor E. Lukacs for his encouragement and for his valuable suggestions and comments. We wish also to express our appreciation to the publishers for their most efficient handling of the publication of this book.

August 1972
Quebec

WALTER HENGARTNER
RADU THEODORESCU

Notations

Abbreviations

N^*	set of natural numbers
N	set of nonnegative integers
R^r	r-dimensional Euclidean space, $R^1 = R$
$\mathscr{B}^r$	σ-algebra of Borel sets in R^r, $\mathscr{B}^1 = \mathscr{B}$
R^r_+	$\{(x_1, \ldots, x_r) \in R^r : x_k > 0, 1 \leq k \leq r\}$, $R^1_+ = R_+$
$f(x \pm 0) = f^{\pm}(x)$	right (left) limit of the function f at x
$\operatorname{int}[x]$	integral part of x
$\operatorname{Re} x$	real part of x
Cl	closure
∂	boundary
tr	trace
$\lvert \boldsymbol{x} \rvert$	$(\sum_{k=1}^{r} x_k^2)^{\frac{1}{2}}$ if $\boldsymbol{x} = (x_1, \ldots, x_r) \in R^r$
$(\boldsymbol{x}, \boldsymbol{y})$	$\sum_{k=1}^{r} x_k y_k$ if $\boldsymbol{x} = (x_1, \ldots, x_r) \in R^r$ and $\boldsymbol{y} = (y_1, \ldots, y_r) \in R^r$
$\{x\}$	set containing only the element x
$*$	convolution either of p.m.'s or d.f.'s
$\circ$	composition of functions
$\times$	Cartesian product
$\varnothing$	empty set

A^c	complement of the set A
diam A	diameter of the set A
χ_A	indicator of A
P	probability
E	expectation either of a r.v., or of a d.f., or of a p.m.
D^2	variance either of a r.v., or of a d.f., or of a p.m.
med	median either of a r.v., or of a d.f., or of a p.m.
$\lambda \ll \mu$	λ absolutely continuous with respect to μ
a.s.	almost surely
c.f.	characteristic function
cn.	concentration
cn.f.	concentration function
d.f.	distribution function
ds.	dispersion
ds.f.	dispersion function
f.	function
f.f.	frequency function
iff	if and only if
i.o.	infinitely often
i.r.v.'s	independent random variables
p.m.	probability measure
q.m.	quadratic mean
r.v.	random variable

1 BASIC PROPERTIES OF LEVY CONCENTRATION FUNCTIONS

Chapter 1 deals with Lévy cn.f.'s. We give here their basic properties. In Section 1.1 these functions are defined and immediate properties listed. Section 1.2 deals with continuity properties. Examples of cn.f.'s are given in Section 1.3. Section 1.4 is concerned with invariance properties; here two important transformations, namely shift and symmetry, are dealt with. Section 1.5 examines the behavior of $Q(l)/l$, where Q is a cn.f.; we start with differentiability properties and then we try to characterize cn.f.'s by means of absolute constants related to $Q(l)/l$. Section 1.6 is devoted to the relationship between cn.f. and variance. Finally, Section 1.7 examines various ways of extending the notion of cn.f. to p.m.'s in r-dimensional Euclidean space.

1.1 DEFINITIONS AND ELEMENTARY PROPERTIES

1.1.1 Consider the measurable space $(R, \mathscr{B})$. We associate with each p.m. μ on $\mathscr{B}$ four real-valued functions on R:

$$Q_{\mu}^{1}(l) = \begin{cases} 0 & \text{if } l \leq 0, \\ \sup\limits_{x \in R} \mu((-l/2, l/2) + x) & \text{if } l > 0, \end{cases}$$

$$Q_\mu^2(l) = \begin{cases} 0 & \text{if } l \leq 0, \\ \sup\limits_{x \in R} \mu([-l/2, l/2) + x) & \text{if } l > 0, \end{cases}$$

$$Q_\mu^3(l) = \begin{cases} 0 & \text{if } l \leq 0, \\ \sup\limits_{x \in R} \mu((-l/2, l/2] + x) & \text{if } l > 0, \end{cases}$$

$$Q_\mu^4(l) = \begin{cases} 0 & \text{if } l < 0, \\ \sup\limits_{x \in R} \mu([-l/2, l/2] + x) & \text{if } l \geq 0, \end{cases}$$

where, in general,

$$A \pm B = \{a \pm b : a \in A, \quad b \in B\}.$$

DEFINITION 1.1.1

The functions Q_μ^i, $1 \leq i \leq 4$, are called *Lévy cn.f.'s* (or, for brevity, *cn.f.'s*).

Clearly these functions are nondecreasing and bounded by 1.

DEFINITION 1.1.2

The inverse functions $Z_\mu^i = (Q_\mu^i)^{-1}$, $1 \leq i \leq 4$, are called *Lévy ds.f.'s* (or, for brevity, *ds.f.'s*).

For example,

$$\begin{aligned} Z^4(p) &= \inf\{l : Q_\mu^4(l) \geq p\} \\ &= \inf\{l : \text{there exists } a \text{ such that } \mu([a, a + l]) \geq p\}; \end{aligned}$$

here $0 \leq p \leq 1$ and the infimum over the empty set is taken as ∞.

The relationship between the cn.f.'s Q_μ^i, $1 \leq i \leq 4$, is described in

PROPOSITION 1.1.3

For every $l > 0$ we have

$$Q_\mu^1(l) = Q_\mu^2(l) = Q_\mu^3(l), \qquad Q_\mu^3(l + 0) = Q_\mu^4(l). \tag{1.1.1}$$

Proof

(a) For every $0 < h < l$ we have

$$Q_\mu^1(l-h) \le Q_\mu^2(l-h) \le Q_\mu^4(l-h) \le Q_\mu^1(l-h/2)$$

and

$$Q_\mu^1(l-h) \le Q_\mu^3(l-h) \le Q_\mu^4(l-h) \le Q_\mu^1(l-h/2).$$

Hence for $h \downarrow 0$ we get

$$Q_\mu^1(l-0) = Q_\mu^2(l-0) = Q_\mu^3(l-0) = Q_\mu^4(l-0).$$

(b) Analogously for every $h > 0$ we have

$$Q_\mu^1(l+h) \le Q_\mu^2(l+h) \le Q_\mu^4(l+h) \le Q_\mu^1(l+2h)$$

and

$$Q_\mu^1(l+h) \le Q_\mu^3(l+h) \le Q_\mu^4(l+h) \le Q_\mu^1(l+2h),$$

so that for $h \downarrow 0$ we get

$$Q_\mu^1(l+0) = Q_\mu^2(l+0) = Q_\mu^3(l+0) = Q_\mu^4(l+0).$$

(c) For every $\varepsilon > 0$ there exists x_ε such that

$$Q_\mu^1(l) < \mu((x_\varepsilon, x_\varepsilon + l)) + \varepsilon$$

and

$$Q_\mu^1(l-h) \ge \mu((x_\varepsilon, x_\varepsilon + l - h))$$

for $0 < h < l$. Thus

$$Q_\mu^1(l) - Q_\mu^1(l-h) \le \mu([x_\varepsilon + l - h, x_\varepsilon + l)) + \varepsilon.$$

Letting $h \downarrow 0$ we obtain

$$Q_\mu^1(l) - Q_\mu^1(l-0) \le \varepsilon$$

for all $\varepsilon > 0$ and hence $Q_\mu^1(l) = Q_\mu^1(l-0)$. Analogously $Q_\mu^2(l) = Q_\mu^2(l-0)$.

(d) Let $\tilde{\mu}$ be the p.m. defined by

$$\tilde{\mu}(A) = \mu(-A), \qquad A \in \mathscr{B}. \tag{1.1.2}$$

Then

$$\mu((x, x + l]) = \tilde{\mu}([-x - l, -x)) = \tilde{\mu}([y, y + l)),$$

and hence $Q_\mu^3(l) = Q_{\tilde{\mu}}^2(l)$, i.e., $Q_\mu^3(l) = Q_\mu^3(l - 0)$.

(e) Let $h_n > 0$, $n \in N^*$; then there are x_n such that

$$-h_n + Q_\mu^4(l + h_n) < \mu([x_n, x_n + l + h_n]).$$

There is also a monotonic subsequence, say $(x_{n_k})_{k \in N^*}$, converging to x_0. It follows that

$$-h_{n_k} + Q_\mu^4(l + h_{n_k}) < \mu([x_0 - |x_{n_k} - x_0|, x_0 + l + h_{n_k} + |x_{n_k} - x_0|])$$

and, letting $h_{n_k} \downarrow 0$, we get

$$Q_\mu^4(l + 0) \leq \mu([x_0, x_0 + l]) = Q^4(l)$$

so that $Q_\mu^4(l + 0) = Q_\mu^4(l)$.

(f) By (a)–(e), we obtain (1.1.1). ◇

As an immediate consequence we get for $l = 0$ the following

COROLLARY

We have

$$(1.1.3) \quad Q_\mu^1(0 + 0) = Q_\mu^2(0 + 0) = Q_\mu^3(0 + 0) = Q_\mu^4(0) = \sup_{x \in R} \mu(\{x\}).$$

Therefore there are only two distinct cn.f.'s, namely $Q_\mu^- \equiv Q_\mu^1 \equiv Q_\mu^2 \equiv Q_\mu^3$ and $Q_\mu^4(l) = Q_\mu(l)$ for any $l \geq 0$. Q_μ^- and Q_μ are very close to one another; hence, any property proved either for Q_μ^- or for Q_μ may be adapted to the other one. Hence, from now on we shall focus our attention on Q_μ.

THEOREM 1.1.4

The cn.f. Q_μ (respectively Q_μ^-) is a right (respectively left) continuous d.f.

Proof

We have seen that Q_μ is nondecreasing and bounded by 1, therefore $Q_\mu(\infty) \le 1$. Since $Q_\mu(2l) \ge \mu([-l, l])$ and since $\lim_{l\to\infty} \mu([-l, l]) = 1$, we have $Q_\mu(\infty) = 1$.

Next, by Proposition 1.1.3, Q_μ is right continuous (respectively Q_μ^- is left continuous). ◇

Proposition 1.1.3 and Theorem 1.1.4 were proved without making use of the d.f. F_μ associated with μ. It follows that F_μ may be defined to be either left continuous or right continuous. In what follows we shall assume that it is right continuous, i.e., $F_\mu(x) = \mu([-\infty, x])$, $x \in R$. On the other hand, Q_μ^- may be also interpreted as the modulus of continuity of F_μ.

Another property of Q_μ is given by

THEOREM 1.1.5

We have $Q_\mu(l) > 0$ for any $l > 0$.

Proof

Suppose the contrary. Then $\mu([x, x + l]) = 0$ for every $x \in R$; hence $\mu \equiv 0$ which leads to a contradiction. ◇

Before proceeding further, let us make some notational conventions. In what follows we shall speak about the cn.f. either of a r.v., or of a d.f., or of a p.m., in the sense that if ξ, F, and μ are respectively a r.v., its d.f., and the p.m. induced by F on $\mathscr{B}$, then we write $Q_\xi = Q_F = Q_\mu$. We shall also use both the symbols $d\mu(x)$ and $dF(x)$ in writing integrals.

1.1.2 Let $\mathfrak{F}$ denote the set of all d.f.'s F on R (i.e., the set of all real-valued nondecreasing functions F, right continuous, and $F(-\infty) = 0$ and $F(+\infty) = 1$), let $\mathfrak{F}_+$ denote the set of all d.f.'s F such that $F^-(0) = 0$, let $\tilde{\mathfrak{F}}_+$ denote the set of all subadditive d.f.'s $F \in \mathfrak{F}_+$, and let $\mathfrak{Q}$ denote the set of all cn.f.'s Q; clearly, by definition and by Theorem 1.1.4, $\mathfrak{Q} \subset \mathfrak{F}_+$. Further, we get

THEOREM 1.1.6

We have $\mathfrak{Q} = \tilde{\tilde{\mathfrak{F}}}_+$. Moreover, $Q_F = F$ for every $F \in \tilde{\tilde{\mathfrak{F}}}_+$.

Proof

From the definition of Q_F it follows

$$Q_F(l'') - Q_F(l') \leq Q_F(l'' - l'), \qquad l' < l'',$$

so that we get $Q_F(l + m) \leq Q_F(l) + Q_F(m)$ for every $l, m \in R$. It follows by Theorem 1.1.4 that $Q \subset \tilde{\tilde{\mathfrak{F}}}_+$.

Conversely, suppose now that $F \in \tilde{\tilde{\mathfrak{F}}}_+$; we have $F(x + l) - F^-(x) \leq F(l)$ for every $x \in R$ and $l > 0$, wherefrom we get $Q_F \leq F$. Further, $Q_F(l) \geq F(x + l) - F^-(x)$ for any $x \in R$, so that for $x = 0$ we have $Q_F \geq F$. Consequently $Q_F = F$. ◇

Thus, the above theorem shows the structure of the set $\mathfrak{Q}$.

1.1.3 Denote by T a mapping from $\mathfrak{F}$ into $\mathfrak{F}$ defined by $T(F) = Q_F$. As an immediate consequence of Theorems 1.1.4 and 1.1.6 we obtain

THEOREM 1.1.7

We have

(a) $T \circ T = T$;

(b) $T\left(\prod_{k=1}^{n} F_k\right) \leq \sum_{k=1}^{n} T(F_k)$ for any $F_k \in \mathfrak{F}$, $1 \leq k \leq n$;

(c) $\max_{l \leq k \leq n} \alpha_k T(F_k) \leq T\left(\sum_{k=1}^{n} a_k F_k\right) \leq \sum_{k=1}^{n} a_k T(F_k)$ for any $F_k \in \mathfrak{F}$, $1 \leq k \leq n$, and for any $a_k \geq 0$, $1 \leq k \leq n$, $\sum_{k=1}^{n} a_k = 1$.

Moreover, the inequalities (b) and (c) are sharp.

Proof

Assertion (a) is a direct consequence of Theorem 1.1.6.

In order to prove (b) we proceed by induction on n. Clearly (b) is valid for $n = 1$. Suppose now (b) valid for n. Then

$$\begin{aligned}
&\prod_{k=1}^{n+1} F_k(x+l) - \prod_{k=1}^{n+1} F_k^-(x) \\
&\qquad = \prod_{k=1}^{n} F_k(x+l)[F_{n+1}(x+l) - F_{n+1}^-(x)] \\
&\qquad\quad + F_{n+1}^-(x)\left[\prod_{k=1}^{n} F_k(x+l) - \prod_{k=1}^{n} F_k^-(x)\right] \\
&\qquad \le (F_{n+1}(x+l) - F_{n+1}^-(x)) + T\left(\prod_{k=1}^{n} F_k\right) \\
&\qquad \le \sum_{k=1}^{n+1} T(F_k).
\end{aligned}$$

Inequality (b) is sharp for

$$F(x) = F_1(x) = F_2(x) = \begin{cases} 0 & \text{if } x < 0, \\ x & \text{if } 0 \le x < 1, \\ 1 & \text{if } 1 \le x. \end{cases}$$

Next, the assertion (c) is obvious and is sharp for $F_k = F$, $1 \le k \le n$. ◇

We get immediately the following

COROLLARY

Let $Q_F = Q_G$. Then

$$\max(\alpha, \beta) Q_F \le Q_{\alpha F + \beta G} \le Q_F,$$

where $\alpha, \beta \ge 0$, $\alpha + \beta = 1$.

From this corollary we deduce that we can decrease the cn.f. by means of convex combinations of d.f.'s having the same cn.f.

1.1.4 Let us discuss the possibility of reaching the upper bound in the definition of the cn.f. In the proof of Proposition 1.1.3 we have already shown that this is feasible. However, we shall give here another proof of this assertion.

THEOREM 1.1.8

Let

$$A_l = \{x : \mu([0, l] + x) = Q_\mu(l)\}, \qquad l \geq 0.$$

Then A_l is nonvoid for every $l \geq 0$. Moreover, A_l is closed for every $l > 0$.

Proof

For $l = 0$ it is sufficient to observe that $A_0 = R$ if μ is nonatomic (cf. Theorem 1.1.3) and

$$A_0 = \{x : \mu(\{x\}) = \text{greatest mass}\}$$

if μ has mass points (cf. corollary of Proposition 1.1.3).

Assume now that $l > 0$ and set

$$A_l^\varepsilon = \{x : \mu([0, l] + x) \geq Q_\mu(l) - \varepsilon\}$$

By definition of the cn.f., $A_l^\varepsilon \neq \varnothing$. Moreover, A_l^ε is compact. Indeed, let $x_n \in A_l^\varepsilon$, $n \in N^*$, and $\lim_{n\to\infty} x_n = x$. Then for every $\eta > 0$ and for $n \geq n_0(\eta)$ we have $|x_n - x| \leq \eta$ and we can write

$$Q_\mu(l) - \varepsilon \leq \mu([0, l] + x_n) \leq \mu([0, l + 2\eta] + x)$$

so that $x \in A_l^\varepsilon$.

Next, let us notice that $A_l^\varepsilon \subset A_l^{\varepsilon'}$ if $\varepsilon < \varepsilon'$ and set $B_l = \bigcap_{\varepsilon > 0} A_l^\varepsilon$. Obviously, $B_l \neq \varnothing$. On the other hand, if $x \in B_l$, then $\mu([0, l] + x) \geq Q_\mu(l) - \varepsilon$ for every $\varepsilon > 0$, so that $A_l \supset B_l$; but $A_l \subset A_l^\varepsilon$ for every $\varepsilon > 0$; hence $A_l \subset B_l$ and we conclude that $A_l = B_l$. ◇

Let us remark also that the set A_l is closed if $l = 0$.

COROLLARY

There exists a compact set K_l such that

$$Q_\mu(l) = \sup_{x \in K_l} \mu([0, l] + x)$$

for every $l > 0$.

The above corollary holds also true for $l = 0$ provided that μ has mass points.

We have seen that for every $l \geq 0$ there is an $x_l \in R$ such that

$$Q_\mu(l) = \mu([0, l] + x_l). \tag{1.1.4}$$

However, this is not true for Q_μ^-. Consider the following two examples:

$$F(x) = \begin{cases} 0 & \text{if } x < 0, \\ \frac{5}{12}x + \frac{1}{4} & \text{if } 0 \leq x < 1, \\ 1 & \text{if } 1 \leq x, \end{cases}$$

$$G(x) = \begin{cases} 0 & \text{if } x < 0, \\ \frac{5}{12}x + \frac{1}{3} & \text{if } 0 \leq x < 1, \\ 1 & \text{if } 1 \leq x. \end{cases}$$

Then there is no x_1 such that either

$$\tfrac{3}{4} = Q_F^1(1) = F^-(x_1 + 1) - F(x_1)$$

or

$$\tfrac{3}{4} = Q_F^3(1) = F(x_1 + 1) - F(x_1)$$

and there is no x_1 such that either

$$\tfrac{3}{4} = Q_G^1(1) = G^-(x_1 + 1) - G(x_1)$$

or

$$\tfrac{3}{4} = Q_G^2(1) = G^-(x_1 + 1) - G^-(x_1).$$

For Q_μ^- we can assert only that there is an x_l such that either $Q_\mu^-(l) = \mu([0, l) + x_l)$, $l > 0$, or $Q_\mu^-(l) = \mu((0, l] + x_l)$, $l > 0$. Clearly, if μ has no mass points, i.e., if F_μ is continuous, then such an x_l exists also for Q_μ^-.

If F_μ is unimodal and the support of μ is R, then x_l is unique. If we drop the assumption that the support of F_μ is R and we add symmetry, then it is always possible to take $x_l = a - l/2$, where a is the unique mode.

Complement

The ds.f. may be useful in the formulation of certain limit theorems. It often provides the right norming constants [Kesten (1972, p. 705)].

Notes and Comments

The cn.f. was defined and used by Lévy (1937) in his famous monograph devoted to sums of r.v.'s [see also Lévy (1954)]. The same function was also defined independently by Littlewood and Offord (1943). Its main properties are to be found in Lévy (1937, 1954), although some of them were already known by Lévy (1937, 1954). Several properties were proved independently by Offord (1943). The structure of the set of cn.f.'s (Theorem 1.1.6) was given by Sâmboan and Theodorescu (1968), and independently by Diaz (1970). Theorem 1.1.8 was proved by several authors, e.g., Lévy (1937, 1954), Tucker (1963), Diaz (1970), and in a more general context by Parthasarathy (1967, p. 61).

1.2 CONTINUITY PROPERTIES

1.1.2 We start with

PROPOSITION 1.2.1

The greatest jump of Q_μ is at 0.

Proof

By Theorem 1.1.7 we have $Q_\mu(l+h) - Q_\mu(l-h) \leq Q_\mu(2h)$ so that letting $h \downarrow 0$ we obtain $Q_\mu(l) - Q_\mu^-(l) \leq Q_\mu(0)$. ◇

Now we can prove

THEOREM 1.2.2

The following three assertions are equivalent: (a) Q_μ is continuous at 0; (b) Q_μ is continuous on R; (c) μ is nonatomic (i.e., F_μ is continuous).

Proof

Clearly (b) implies (a); by Proposition 1.2.1 it follows that (a) implies (b). Therefore, (a) and (b) are equivalent. Next, by (1.1.3) it follows that (a) and (c) are equivalent. ◇

1.2.2 The following property concerns discrete d.f.'s with a finite set of discontinuities.

THEOREM 1.2.3

If F is a discrete d.f. with a finite set of discontinuities, then Q_F is also of the same kind.

Proof

Let F be a discrete d.f. with a finite set of discontinuities, and let us set

(1.2.1) $$\{y : y = F(x+l) - F^-(x),\ x \in R,\ l \geqslant 0\}.$$

Clearly (1.2.1) is a finite set, containing at most 2^m elements, where m is the number of discontinuities of F. Therefore Q_F is also discrete with a finite set of discontinuities. ◇

The converse of Theorem 1.2.3 is not true. In fact for

$$F(x) = \begin{cases} 0 & \text{if } x < 0, \\ 0.1 & \text{if } 0 \le x < 1, \\ 0.1x & \text{if } 1 \le x < 2, \\ 0.2 & \text{if } 2 \le x < 4, \\ 0.8 & \text{if } 4 \le x < 6, \\ 0.9 & \text{if } 6 \le x < 8, \\ 1 & \text{if } 8 \le x, \end{cases} \tag{1.2.2}$$

we get

$$Q_F(l) = \begin{cases} 0 & \text{if } l < 0, \\ 0.6 & \text{if } 0 \le l < 2, \\ 0.7 & \text{if } 2 \le l < 4, \\ 0.8 & \text{if } 4 \le l < 6, \\ 0.9 & \text{if } 6 \le l < 8, \\ 1 & \text{if } 8 \le l. \end{cases}$$

Notes and Comments

Continuity properties are mainly due to Lévy (1937, 1954), e.g., Theorem 1.2.2. Theorem 1.2.3 is to be found is Sâmboan and Theodorescu (1958).

1.3 EXAMPLES

1.3.1 We start with discrete d.f.'s having a finite set of discontinuities.

(1) If $F = H_a$, then we have $Q_{H_a} = H$, where H_a is the Heaviside d.f. with mass point at x, $H_0 = H$.

(2) If $F = F(\cdot\,; c) = cH_{x_0} + (1-c)H_{x_1}$ with $0 \le c \le 1$, $x_0 < x_1$, then we have

$$Q_{F(\cdot\,;c)}(l) = \begin{cases} 0 & \text{if } l < 0, \\ \max(c, 1-c) & \text{if } 0 \le l < x_1 - x_0, \\ 1 & \text{if } x_1 - x_0 \le l, \end{cases}$$

i.e.,

$$Q_{F(\cdot\,;c)} = \max(c, 1-c)H_0 + (1 - \max(c, 1-c))H_{x_1 - x_0}.$$

Moreover, $Q_{F(\cdot\,;c)} = Q_{F(\cdot\,;1-c)}$, and

$$Q_{F(\cdot\,;c)} = \begin{cases} F(\cdot\,; c) & \text{if } c \le \frac{1}{2}, \\ F(\cdot\,; 1-c) & \text{if } c \ge \frac{1}{2}. \end{cases}$$

(3) If $x_0 \le x_1 \le \cdots \le x_n$, $|x_i| \le M$, $1 \le i \le n$, and if

$$F(x) = \begin{cases} 0 & \text{if } x < x_0, \\ i/n & \text{if } x_{i-1} \le x < x_i, \quad 1 \le i \le n, \end{cases}$$

then we have

$$Q_F(l) = \begin{cases} 0 & \text{if } l < 0, \\ i/n & \text{if } x'_{i-1} < l \le x'_i, \quad 1 \le i \le n; \end{cases}$$

here the x'_i belong to the set $\{|x_i - x_j| : 0 \le i < j \le n\}$. Moreover, if $x_0 = 0$ and the x_i are equidistant, then $F \in \tilde{\mathfrak{F}}_+$ and by Theorem 1.1.7, $Q_F = F$ In this case $x'_i = x_i$, $0 \le i \le n$.

If we consider n d.f.'s F_i, $1 \le i \le n$, where

$$F_i(x) = \begin{cases} 0 & \text{if } x < -a_i, \\ \frac{1}{2} & \text{if } -a_i \le x < a_i, \\ 1 & \text{if } a_i \le x \end{cases}$$

provided that all the a_i are positive and

$$\sum_{i=1}^{r} \varepsilon_i a_i \ne 0, \qquad \varepsilon_i = \pm 1, \quad 1 \le i \le r, \quad 1 \le r \le n, \tag{1.3.1}$$

then the d.f. $F = \mathop{\ast}\limits_{i=1}^{n} F_i$ has 2^n discontinuities with jumps all equal to 2^{-n}. Hence, it is of the form considered above. Moreover, these discontinuities are symmetrically distributed about 0. If we drop condition

(1.3.1), then F has at most 2^n discontinuities with jumps of the form $m2^{-n}$, $1 \leq m \leq 2^n$.

(4) If $F = \sum_{i \in I} c_i H_{x_i}$, where $0 \leq c_i \leq 1$, $i \in I$, $\sum_{i \in I} c_i = 1$, $x_{i-1} < x_i$, $i \in I$, I being finite, then for $F \in \mathfrak{F}_+$ clearly $Q_F = F$. If $F \notin \mathfrak{F}_+$, then the new mass points x_i' belong to the set $\{|x_i - x_j| : i < j, i, j \in I\}$. At the first mass point, i.e. at 0, Q_F has the greatest jump, and $Q_F(0) = c_1' = \max_{i \in I} c_i$.

1.3.2 Let us consider now absolutely continuous d.f.'s.

(1) If $F(\cdot) = \Phi(m, \sigma^2; \cdot)$, where $\Phi(m, \sigma^2; \cdot)$ is the normal d.f. with expectation m and variance σ^2, then we have

$$\begin{aligned} Q_F(l) &= \Phi(m, \sigma^2; m + l/2) - \Phi(m, \sigma^2; m - l/2) \\ &= 2\Phi(m, \sigma^2; m + l/2) - 1 \\ &= 2\Phi(0, \sigma^2; l/2) - 1, \qquad l \geq 0, \end{aligned}$$

and the corresponding f.f. is $\phi(0, \sigma^2; l/2)$, $l \geq 0$, where $\phi(m, \sigma^2; \cdot)$ is the f.f. corresponding to $\Phi(m, \sigma^2; \cdot)$.

We notice that Q_F does not depend on m (see p. 15).

(2) If $F = U_{[a, b]}$, where $U_{[a, b]}$ is the uniform d.f. on the bounded interval $[a, b]$, then we have

$$Q_F(l) = \begin{cases} 0 & \text{if } l < 0, \\ (b - a)^{-1} l & \text{if } 0 \leq l < b - a, \\ 1 & \text{if } b - a \leq l. \end{cases}$$

(3) If $F(\cdot) = C(\cdot; \alpha, \beta)$ is the Cauchy d.f. with parameters $\alpha > 0$ and $\beta \in R$ defined by the f.f. $c(x; \alpha, \beta) = \pi^{-1}\alpha(\alpha^2 + (x - \beta)^2)^{-2}$ for every $x \in R$, then we have

$$Q_F(l) = 2\pi^{-1} \arctan(l/2\alpha)$$

for every $l > 0$. Q_F being an increasing function, its inverse (i.e., the ds.f.) is given by $p = 2\pi^{-1} \arctan(l/2\alpha)$, that is $l = Q_\mu^{-1}(p) = 2\alpha \tan(\pi p/2)$. For $p = 1$, we get $l = \infty$; for $p = 0$, we get $l = 0$; for $p = \frac{1}{2}$, we get $l = 2\alpha$. Consequently, α is the abscissa of the third quartile.

Notes and Comments

The examples discussed are to be found in Sâmboan and Theodorescu (1968) and in Dacuhna-Castelle, Revuz, and Schreiber (1970).

1.4 INVARIANCE PROPERTIES

1.4.1 Consider a *shift* by an amount a of μ, i.e.,

$$\mu_a(A) = \mu(A - a) \tag{1.4.1}$$

for every $A \in \mathscr{B}$. In other words, $\mu_a = \mu * \delta_a$, where δ_a is the p.m. with unique mass point at a, and $\delta_0 = \delta$.

THEOREM 1.4.1

We have $Q_{\mu_a} = Q_\mu$.

Proof

In fact

$$\begin{aligned} Q_{\mu_a}(l) &= \sup_{x \in R} \mu_a([-l/2, l/2] + x) \\ &= \sup_{x+a \in R} \mu([-l/2, l/2] + x+a) = Q_\mu(l). \quad \diamondsuit \end{aligned}$$

We conclude that the cn.f. is invariant with respect to the operation $\mu \to \mu_a$.

More generally, we have

THEOREM 1.4.2

We have $Q_{\mu_a * \nu_b} = Q_{\mu * \nu}$.

Proof

We can write

$$\begin{aligned} Q_{\mu_a * \nu_b} &= Q_{\mu * \delta_a * \nu_b} \\ &= Q_{(\mu * \nu_b) * \delta_a} \\ &= Q_{\mu * \nu_b} = Q_{(\mu * \nu) * \delta_b} = Q_{\mu * \nu}. \quad \diamond \end{aligned}$$

The following result concerns the problem of the invariance of the cn.f. with respect to the operation $\mu \to \tilde{\mu}$ defined by (1.1.2).

THEOREM 1.4.3

We have $Q_{\tilde{\mu}} = Q_{\mu}$.

Proof

We have seen that part (d) of the proof of Proposition 1.1.3 shows that $Q_{\mu}^{3} = Q_{\tilde{\mu}}$. Since $Q_{\mu}^{3} = Q_{\mu}$, we conclude that $Q_{\tilde{\mu}} = Q_{\mu}$. $\diamond$

The cn.f. is thus invariant under the operations (1.1.2) and (1.4.1) on p.m.'s. In other words, if for $F \in \mathfrak{F}$ we set

$$F_a(x) = F(x - a), \qquad x \in R, \tag{1.4.2}$$

and

$$\tilde{F}(x) = 1 - F(-x - 0) \qquad x \in R, \tag{1.4.3}$$

then F_a and $\tilde{F}$ have the same cn.f. as F, i.e., the cn.f. is invariant under the operations (1.4.2) and (1.4.3) on d.f.'s. This fact clarifies why for the d.f. considered on p. 13, $F(\cdot; c) = cH_{x_0} + (1 - c)H_{x_1}$, we have $Q_{F(\cdot\,;c)} = Q_{F(\cdot\,;1-c)}$.

The converse is not true, in the sense that $Q_F = Q_G$, and G is neither F_a nor $\tilde{F}$ nor a mixture of them, as shown by the following example. Let

$$F(x) = \begin{cases} 0 & \text{if } \ x < 1, \\ 0.40 & \text{if } \ 1 \le x < 2, \\ 0.65 & \text{if } \ 2 \le x < 2.5, \\ 1 & \text{if } \ 2.5 \le x; \end{cases}$$

we have

$$Q_F(l) = \begin{cases} 0 & \text{if } \ l < 0, \\ 0.40 & \text{if } \ 0 \le l < 0.50, \\ 0.60 & \text{if } \ 0.50 \le l < 1.00, \\ 0.65 & \text{if } \ 1.00 \le l < 1.50, \\ 1 & \text{if } \ 1.50 \le l. \end{cases}$$

The d.f. F cannot be obtained from Q_F by the operations (1.4.2) or (1.4.3); nevertheless their cn.f.'s coincide.

1.4.2 The example above shows that Q_F may have more discontinuities than F. However, it may also happen that the number of discontinuities of Q_F is less than the number of discontinuities of F. For an illustration, modify the example (1.2.2); take

$$F(x) = \begin{cases} 0 & \text{if } \ x < 0, \\ 0.1 & \text{if } \ 0 \le x < 1, \\ F_1(x) & \text{if } \ 1 \le x < 2, \\ 0.2 & \text{if } \ 2 \le x < 4, \\ 0.8 & \text{if } \ 4 \le x < 6, \\ 0.9 & \text{if } \ 6 \le x < 8, \\ 1 & \text{if } \ 8 \le x, \end{cases} \tag{1.4.4}$$

where F_1 is an arbitrary step function chosen so that F is a d.f.; then

$$Q_F(l) = \begin{cases} 0 & \text{if } \ l < 0, \\ 0.6 & \text{if } \ 0 \le l < 2, \\ 0.7 & \text{if } \ 2 \le l < 4, \\ 0.8 & \text{if } \ 4 \le l < 6, \\ 0.9 & \text{if } \ 6 \le l < 8, \\ 1 & \text{if } \ 8 \le l. \end{cases}$$

We can only assert that if F is a discrete d.f. with a finite set of discontinuities, then Q_F is also of that kind (cf. Theorem 1.2.3).

1.4.3 Suppose that in example (1.4.4) F_1, otherwise arbitrary, is chosen so that F is a d.f. In this case, this modified example shows that we cannot characterize the class of d.f.'s leading to the same cn.f. However, under additional restrictions, special classes may be isolated.

THEOREM 1.4.4

Let F and G be absolutely continuous, symmetric, and unimodal, and suppose that $Q_F = Q_G$. Then there is $a \in R$ such that $F = G_a$.

Proof

There are x_l^F and x_l^G such that $Q_F(l) = F(x_l^F + l) - F(x_l^F)$ and $Q_G(l) = G(x_l^G + l) - G(x_l^G)$, so that we can write

$$\begin{aligned} F(x_l^F + l) - F(x_l^F) &= G(x_l^G + l) - G(x_l^G) \\ &= G(x_l^F + \varepsilon_l + l) - G(x_l^F + \varepsilon_l), \end{aligned}$$

where $x_l^F = a^F - l/2$, $x_l^G = a^G - l/2$. Here a^F and a^G are the modes of F and G respectively. It follows that

$$F(a^F + l/2) - F(a^F - l/2) = G(a^F + l/2 - a) - G(a^F - l/2 - a),$$

where $a = a^F - a^G$. Consequently,

$$F(a^F + l/2) - F(a^F - l/2) = G_a(a^F + l/2) - G_a(a^F - l/2)$$

with $G_a(x) = G(x - a)$. Hence $F \equiv G_a$. ◇

Notes and Comments

Invariance of the cn.f. with respect to shifts goes back to Lévy (1937, 1954). The same property with respect to symmetries and Theorem 1.1.9 are due to Hengartner and Theodorescu (1972a).

1.5 BEHAVIOR OF $Q(l)/l$

1.5.1 We start with the differentiability of the subadditive function $Q \in \mathfrak{Q}$ at $l = 0$. Let us set

$$A = \inf_{l<0} Q(l)/l, \qquad B = \sup_{l>0} Q(l)/l.$$

If A and B are finite, then by Hille and Philipps (1957, p. 250),

$$A = \lim_{l \uparrow 0} Q(l)/l, \qquad B = \lim_{l \downarrow 0} Q(l)/l \tag{1.5.1}$$

and $0 \leq A \leq B$; the same conclusion is valid for $B = +\infty$ provided that $\lim_{l \to 0} Q(l) = 0$. Moreover (*ibidem*, p. 251), at each l where Q is differentiable we have $0 = A \leq Q'(l-0),\ Q'(l+0) \leq B$.

THEOREM 1.5.1

Q is not differentiable at $l = 0$.

Proof

It suffices to assume Q continuous at $l = 0$. We have by the subadditivity of Q

$$0 \leq (Q(l+h) - Q(l))/h \leq Q(h)/h, \qquad h > 0.$$

Suppose now that Q is differentiable at the origin; then it follows that Q is differentiable on R and $Q' \equiv 0$ which leads to a contradiction.

1.5.2 Suppose that $F \in \mathfrak{F}_+$ and let K be a constant such that

$$F(X) \leq K(X/x)F(x) \tag{1.5.2}$$

for any $X \geq x > 0$. Obviously, $K \geq 1$. In what follows we shall denote by K_F the smallest of the K satisfying (1.5.2). It is clear that

$$\begin{aligned} K_F &= \inf\{K : F(ax) \leq KaF(x), \quad a \geq 1, \quad x > 0\} \\ &= \sup_{\substack{x>0 \\ a \geq 1}} \frac{F(ax)}{aF(x)} = \sup_{X \geq x > 0} \frac{F(X)/X}{F(x)/x}. \end{aligned}$$

PROPOSITION 1.5.2

If $F \in \mathfrak{F}_+$, then $K_{Q_F} \leq K_F$.

Proof

For all $a \geq 1$, we can write

$$\begin{aligned} Q_F(al) &= \sup_{x \in R} (F(x + al) - F^-(x)) = \sup_{x \in R} (F(a(x/a + l)) - F^-(ax/a)) \\ &\leq \sup_{x \in R} [K_F a(F(x/a + l) - F(x/a))] = K_F a Q_F(l), \end{aligned}$$

so that $K_{Q_F} \leq K_F$. ◇

Further, we have

THEOREM 1.5.3

Let $Q \in \mathfrak{Q}$. Then $1 \leq K_Q \leq 2$. Moreover, for each $r \in [1, 2]$ there is a $Q \in \mathfrak{Q}$ with $K_Q = r$.

Proof

Let $L \geq l > 0$; since Q is subadditive, we have

$$\begin{aligned} Q(L) = Q((L/l)l) &\leq Q((\text{int}[L/l] + 1)l) \\ &\leq (\text{int}[L/l] + 1)Q(l) \leq 2(L/l)Q(l). \end{aligned}$$

Now, take

$$F(x) = \begin{cases} 0 & \text{if } \ x < 0, \\ 1/r & \text{if } \ 0 \leq x < x_0, \\ 1 & \text{if } \ x_0 \leq x; \end{cases}$$

then $Q_F = F$ and $K_{Q_F} = r$. ◇

On the other hand, no absolute upper bound is available for K_F, since for every $r \in [1, \infty)$ we can construct an absolutely continuous d.f. F such that $K_F = r$; for example

$$F(x) = \begin{cases} 0 & \text{if } x < 0, \\ x & \text{if } 0 \le x < \varepsilon, \\ \varepsilon & \text{if } \varepsilon \le x < \frac{1}{2}, \\ (2 - 2\varepsilon)(x - 1) + 1 & \text{if } \frac{1}{2} \le x < 1, \\ 1 & \text{if } 1 \le x. \end{cases}$$

Let us give now an application of Theorem 1.5.3.

PROPOSITION 1.5.4

Let $F, G \in \mathfrak{F}$. Then there exists a constant A_G, depending on G but independent of F, such that

$$\sup_{x \in R} |F * G(x) - F(x)| \le A_G\, Q_F(\sigma_G), \tag{1.5.3}$$

where $\sigma_G^2 > 0$ is the variance of G. Moreover, A_G may be replaced by an absolute constant A if the expectation of G is 0, independent of σ_G.

Proof

If $\sigma_G = \infty$, then $\sup_{x \in R} |F * G(x) - F(x)| \le 1$. Suppose now $0 < \sigma_G < \infty$. Then

(1.5.4)

$$\sup_{x \in R} |F * G(x) - F(x)| = \sup_{x \in R} \left| \int_R (F(x - u) - F(x))\, dG(u) \right|$$

$$\le \int_R Q_F(|u|)\, dG(u).$$

But

$$Q_F(|u|) \le \begin{cases} Q_F(\sigma_G) & \text{if } |u| \le \sigma_G, \\ 2(|u|/\sigma_G) Q_F(\sigma_G) & \text{if } |u| > \sigma_G, \end{cases} \tag{1.5.5}$$

by Theorem 1.5.3. Taking into account (1.5.5) in (1.5.4), we get

$$\sup_{x \in R} |F * G(x) - F(x)| \le Q_F(\sigma_G)\left(1 + (2/\sigma_G) \int |u|\, dG(u)\right);$$

hence

$$A_G = 1 + (2/\sigma_G) \int_R |u| \, dG(u).$$

Suppose now that the expectation of G is 0. If $\sigma_G = 0$, (1.5.3) is trivial; hence, consider $\sigma_G > 0$ and in this case $A_G \leq 3$, so that $A = 3$. ◇

In Theorems 1.5.5–1.5.8 we try to characterize Q (respectively F) by means of K_Q (respectively K_F).

THEOREM 1.5.5

If $Q \in \mathfrak{Q}$ and $K_Q = 1$, then $Q = cH_0 + (1-c)Q_1$, where Q_1 is absolutely continuous.

Proof

We shall show first that Q is absolutely continuous on $I_m = [1/m, \infty)$, $m \in N^*$. Let $E \subset I_m$ be the union of disjoint intervals

$$E = \bigcup_{k=1}^{n} [x_k, x_k + h_k].$$

Then

$$\sum_{k=1}^{n} (Q(x_k + h_k) - Q(x_k)) \leq \sum_{k=1}^{n} (Q(x_k)/x_k)h_k \leq m \sum_{k=1}^{n} h_k < \varepsilon$$

provided that $\sum_{k=1}^{n} h_k < \delta = \varepsilon/m$. Hence, by the Lebesgue monotone convergence theorem there is an integrable function on R, say f, such that

$$Q(l) - Q(0) = \int_{(0,l]} f(u) \, du,$$

i.e., $Q(l) = cH_0 + (1-c)Q_1$. This follows from the fact that, for the p.m. μ_Q induced by Q, we have

$$\mu_Q([a, b]) = \mu_Q([a, b] \cap \{0\}) + \mu_Q([a, b] \cap (0, \infty))$$

for every $0 \leq a < b$. ◇

The geometrical interpretation of the condition $K_Q = 1$ is that the set $D = \{(l, m) : l > 0, 0 < m < Q(l)\}$ is starlike with respect to (0, 0).

COROLLARY

If $Q \in \mathfrak{Q}$, $K_Q = 1$ and $Q(0) = 0$, then Q is absolutely continuous.

THEOREM 1.5.6

Let $F \in \mathfrak{F}_+$ and $K_F = 1$. Then $Q_F = F$ and $K_Q = 1$.

Proof

It suffices to prove that F is subadditive. For $y \geqslant x > 0$, we have

$$F(x + y) \leq ((x + y)/y)F(y) = F(y) + (x/y)F(y) \leq F(x) + F(y);$$

hence $F \in \tilde{\mathfrak{F}}_+$ and consequently $Q_F = F$. ◇

THEOREM 1.5.7

Let $r \in [1, 2)$. Then there is a $Q \in \mathfrak{Q}$, absolutely continuous such that $r = K_Q$ and its f.f. is bounded by 1.

Proof

Let $Q \in \mathfrak{Q}$ be defined by

$$Q(l) = \begin{cases} 0 & \text{if } l < 0, \\ l & \text{if } 0 \leq l < \frac{1}{2}, \\ \frac{1}{2} & \text{if } \frac{1}{2} \leq l < 1/(2 - r) - \frac{1}{2}, \\ l + 1 - 1/(2 - r) & \text{if } 1/(2 - r) - \frac{1}{2} \leq l < 1/(2 - r), \\ 1 & \text{if } 1/(2 - r) \leq l. \end{cases}$$

It is easily seen that $K_Q = r$. ◇

THEOREM 1.5.8

Suppose $Q(0) > 0$. Then $K_Q = 2$ iff Q has the form

$$Q(l) = \begin{cases} 0 & \text{if } \; l < 0, \\ a \le \frac{1}{2} & \text{if } \; 0 \le l < x_0, \\ 2a & \text{if } \; x_0 \le l < 2x_0, \\ Q_1(l) & \text{if } \; 2x_0 \le l, \end{cases}$$

where $Q_1(2x_0) \ge 2a$.

Proof

Let us take $L = x_0 + \varepsilon$. Then

$$2a = Q(x_0 + \varepsilon) \le K_Q((x_0 + \varepsilon)/x_0)a = K_Q a(1 + \varepsilon/x_0)$$

and $K_Q \ge Q(x_0 + \varepsilon)/a(1 + \varepsilon/x_0)$. Letting $\varepsilon \downarrow 0$, we get $K_Q \ge 2$, which implies that $K_Q = 2$.

Conversely, let us assume that $K_Q = 2$. For every $n \in N^*$ there exist $y_n \ge x_n > 0$ such that

$$2(y_n/x_n)Q(x_n) \le Q(y_n) + 1/n \le 2(y_n/x_n)Q(x_n) + 1/n.$$

Let $y_n = \alpha_n x_n + \beta_n$, $n \in N^*$, where $\alpha_n \in N^*$ and $\beta_n \in [0, x_n)$. Then

$$(1.5.6) \qquad 2\alpha_n Q(x_n) + 2(\beta_n/x_n)Q(x_n) \le \alpha_n Q(x_n) + Q(\beta_n) + 1/n$$

and

$$\alpha_n Q(x_n) \le Q(\beta_n) - 2(\beta_n/x_n)Q(x_n) + 1/n \le Q(x_n) + 1/n$$

so that $\alpha_n \le 1$ for $n \ge n_0$, i.e., $\alpha_n = 1$ for $n \ge n_0$. With no less generality we may assume that $y_n = x_n + \beta_n$, $n \in N^*$.

It follows from (1.5.6) that

$$2Q(x_n) \le Q(x_n) + Q(\beta_n) + 1/n - 2(\beta_n/x_n)Q(x_n)$$

so that

$$\begin{aligned} Q(x_n) &\le Q(\beta_n) - 2(\beta_n/x_n)Q(x_n) + 1/n \\ &\le Q(x_n)(1 - 2(\beta_n/x_n)) + 1/n, \end{aligned}$$

which implies that $0 \le Q(x_n)2(\beta_n/x_n) \le 1/n$. Suppose that $Q(x_n)/x_n$ does not tend to 0 as $n \to \infty$. Then $\beta_n \to 0$ as $n \to \infty$. This is clearly the case if $\liminf_{x \downarrow 0} Q(x)/x > 0$ since we cannot have $x_n \to \infty$ as $n \to \infty$. In fact if $x_n \to \infty$ as $n \to \infty$, the inequality

$$2Q(x_n) \le 2((x_n + \beta_n)/x_n)Q(x_n) \le Q(2x_n) + 1/n$$

would be violated for $Q(x_n) > \frac{3}{4}$ and $n > 4$. Since $\sup_{l>0} Q(l)/l = B > 0$ it follows from (1.5.1) that $\lim_{l \downarrow 0} Q(l)/l = B > 0$. From the inequalities

$$\begin{aligned} Q(0) \le \liminf_{n \to \infty} Q(x_n) &\le \limsup_{n \to \infty} Q(x_n) \\ &\le \limsup_{n \to \infty} Q(x_n)(1 + 2(\beta_n/x_n)) \\ &\le \limsup_{n \to \infty} Q(\beta_n) + 1/n = Q(0) \end{aligned}$$

we conclude that $\lim_{n \to \infty} Q(x_n) = Q(0)$.

There are two possible cases:

(a) There exists a subsequence $(x_{n_k})_{k \in N^*}$ converging to $x_0 \ne 0$ as $k \to \infty$. Then Q has a discontinuity at x_0 of height $Q(0)$ and hence $Q(0) \le \frac{1}{2}$. Finally, subadditivity of Q implies that

$$Q(l) = \begin{cases} 0 & \text{if } l < 0, \\ a = Q(x_0) & \text{if } 0 \le l < x_0, \\ 2a & \text{if } x_0 \le l < 2x_0, \\ Q_1(l) & \text{if } 2x_0 \le l, \end{cases}$$

with $Q_1(2x_0) \ge 2a$.

(b) If $x_n \to 0$ as $n \to \infty$, then there exists a subsequence $(n_k)_{k \in N^*}$ such that $x_{n_{k+1}} \le x_{n_{k+1}} + \beta_{n_{k+1}} \le x_{n_k}$. Hence

$$2Q(0) \le 2Q(x_{n_{k+1}}) \le Q(x_{n_k}) + 1/n_k.$$

Now choose $j_0 > 0$ such that $2^{j_0}a > 2$ and a subsequence $(n_{k_j})_{j \in N^*}$ such that $\sum_{j=1}^{j_0} (2^{j-1}/n_{k_j}) < 1$. Then Q cannot be bounded by 1 and case (b) is excluded. ◇

Complements

1 Theorem 1.5.3 may be restated also in the following form. Let $Q \in \mathfrak{Q}$; if $\alpha > 0$, then $Q(\alpha l) \leq (\mathrm{int}[\alpha] + 1)Q(l)$ [Esséen (1968)].

2 Let $F \in \mathfrak{F}_+$ and let A^α be a constant such that

$$F(X) \leq A^\alpha (X/x)(1 - F(x))^{-\alpha}, \qquad \alpha \geq 0, \tag{1.5.7}$$

for all $X \geq x > 0$. In what follows we shall denote by A_F^α the smallest of the A^α with the above property. Clearly

$$\begin{aligned} A_F^\alpha &= \inf\{A^\alpha : F(ax) \leq A^\alpha a(1 - F(x))^{-\alpha}, \quad x > 0, \quad a \geq 1\} \\ &= \sup_{\substack{x>0 \\ a \geq 1}} \frac{F(ax)}{a(1 - F(x))^{-\alpha}} = \sup_{X \geq x > 0} \frac{F(X)/X}{(1 - F(x))^{-\alpha}/x}. \end{aligned}$$

Moreover

$$\alpha^\alpha (1 + \alpha)^{-1-\alpha} \leq A_F^\alpha \leq K_F \alpha^\alpha (1 + \alpha)^{-1-\alpha}; \tag{1.5.8}$$

for $Q \in \mathfrak{Q}$ we have $K_Q \in [1, 2]$ in (1.5.8). Cn.f.'s may be also characterized by A_Q^α. Note also that (1.5.7) may be considered as a Kolmogorov type inequality (see Section 2.2).

Notes and Comments

The results concerning the behaviour of $Q(l)/l$ are due to Sâmboan and Theodorescu (1968) and to Hengartner and Theodorescu (1972a). Proposition 1.5.4 was stated without proof by Kolmogorov (1956) as an auxiliary result in proving some limit theorems.

1.6 RELATIONSHIP BETWEEN CONCENTRATION FUNCTION AND VARIANCE

1.6.1 We start with an elementary result.

PROPOSITION 1.6.1

Let μ be a p.m. and suppose that its variance σ_μ^2 is finite. Then

$$Q_\mu(2l) \geq 1 - \sigma_\mu^2/l^2 \tag{1.6.1}$$

for every $l > 0$.

Proof

(1.6.1) is a straightforward consequence of the Tchebychev inequality. ◇

If we take in (1.6.1) $l = \lambda\sigma_\mu$ with $\lambda > 1$, then $Q_\mu(2\lambda\sigma_\mu) \geq 1 - 1/\lambda^2$. We deduce also from (1.6.1) that $l^2(1 - Q_\mu(l)) \leq 4\sigma_\mu^2$.

1.6.2 Let $0 < \alpha < 1$ and $l > 0$ be given. Set

$$\mathfrak{M}_1 = \{\mu : Q_\mu(2l) \leq \alpha\}.$$

Since by the Tchebychev inequality $1 - \sigma_\mu^2/l^2 \leq Q_\mu(2l) \leq \alpha$, it is natural to try to find $\inf_{\mu \in \mathfrak{M}_1} \sigma_\mu^2$. We get immediately a lower bound of σ_μ^2, namely $\sigma_\mu^2 \geq l^2(1 - \alpha)$. Now we can prove

THEOREM 1.6.2

Let $0 < \alpha \leq 1$ and $l > 0$ be given. Then

$$\sigma_\mu^2 \geq (l^2/3)p(p + 1)(3 - \alpha(2p + 1))$$

and equality is reached for $\mu_0 = \sum_{k=0}^{2p} \beta_k \delta_{(k-p)l}$, where $1/(p + 1) \leq \alpha < 1/p$, $p \in N^*$, or $\alpha = 1$ and

$$\beta_k = \begin{cases} 1 - p\alpha & \text{if } k \text{ is even,} \\ (p + 1)\alpha - 1 & \text{if } k \text{ is odd.} \end{cases}$$

Proof

If $\alpha = 1$ then $\inf_{\mu \in \mathfrak{M}_1} \sigma_\mu^2 = 0$ and $\mu_0 = \delta_0$. Let now $\alpha < 1$. Then there exists $p \in N^*$ such that $1/(p + 1) \leq \alpha < 1/p$. Suppose $\varepsilon > 0$ given; then for each $\mu \in \mathfrak{M}_1$ there exists a nonatomic $\mu_1 \in \mathfrak{M}_1$ such that $\sigma_{\mu_1}^2 \leq \sigma_\mu^2 + \varepsilon$. First we show that if $\mu_1 \in \mathfrak{M}_1$ is nonatomic, there is an atomic p.m. $\mu_2 = \sum_{k=0}^{2p} \beta_k \delta_{z_k} \in \mathfrak{M}_1$, where

$$\beta_k = \begin{cases} 1 - p\alpha & \text{if } k \text{ is even,} \\ (p+1)\alpha - 1 & \text{if } k \text{ is odd,} \end{cases}$$

such that $\sigma_{\mu_2}^2 \leq \sigma_{\mu_1}^2$. In fact, there exists a partition $-\infty = x_0 < x_1 < \cdots < x_{2p+1} = \infty$ such that $\beta_k = \mu_1((x_k, x_{k+1}])$, $0 \leq k \leq 2p$. If $\alpha = 1/(p+1)$, we put $x_{2k+1} = x_{2k+2}$, $0 \leq k \leq p-1$. Let z_k, $1 \leq k \leq 2p$, be the centers of gravity for μ_1 restricted to the interval (x_k, x_{k+1}). If $\alpha = 1/(p+1)$, then $z_{2k+1} = x_{2k+1}$, $0 \leq k \leq p-1$. Put $\mu_2 = \sum_{k=0}^{2p} \beta_k \delta_{z_k}$; clearly $\sigma_{\mu_2}^2 \leq \sigma_{\mu_1}^2 \leq \sigma_\mu^2 + \varepsilon$. We shall show that $\mu_2 \in \mathfrak{M}_1$. Since $\beta_k + \beta_{k+1} = \alpha$, it suffices to prove that $z_{k+2} - z_k \geq 2l$, $0 \leq k \leq p-2$.

(a) If $\beta_k = 0$, then $z_{k+2} - z_k = x_{k+2} - x_{2k}$ and $\mu_1((x_k, x_{k+2}]) = \alpha$ implies $z_{2k+2} - z_k \geq 2l$.

(b) If $\beta_k > 0$, then

$$\begin{aligned} z_{k+2} - z_k &= \beta_{k+2}^{-1} \int_{x_{k+2}}^{x_{k+3}} x \, d\mu_1(x) - \beta_k^{-1} \int_{x_k}^{x_{k+1}} x \, d\mu_1(x) \\ &= \beta_k^{-1} \left(\int_{x_{k+2}}^{x_{k+3}} x \, d\mu_1(x) - \int_{x_k}^{x_{k+1}} x \, d\mu_1(x) \right). \end{aligned}$$

Put now $y = F(x) = \mu_1((-\infty, x])$; then

$$z_{k+2} = \beta_k^{-1} \int_{F(x_{k+2})}^{F(x_{k+3})} F^{-1}(y) \, dy = \beta_k^{-1} \int_{F(x_k)}^{F(x_{k+1})} F^{-1}(y + \alpha) \, dy.$$

Therefore

$$z_{k+2} - z_k = \beta_k^{-1} \int_{\mu_1((-\infty, x_k])}^{\mu_1((-\infty, x_{k+1}])} (F^{-1}(y + \alpha) - F^{-1}(y)) \, dy \geq 2l,$$

i.e., $\mu_2 \in \mathfrak{M}_1$.

Choose $z'_p = 0$ and $z'_{p+1} \in (0, 2l)$ and take $z'_{p+2k} = 2kl$ for every $-p \leq 2k \leq p$ and $z'_{p+2k+1} = z'_{p+1} + 2kl$ for every $-p \leq 2k+1 \leq p$.

Put

$$\mathfrak{M}_2 = \left\{\mu : \mu = \sum_{k=0}^{2p} \beta_k \, \delta_{z_{k'}}\right\}.$$

Then $\mathfrak{M}_2 \subset \mathfrak{M}_1$ and there is a $\mu_3 \in \mathfrak{M}_2$ such that

$$\sigma_{\mu_3}^2 \leq \sigma_{\mu_2}^2 \leq \sigma_{\mu_1}^2 \leq \sigma_\mu^2 + \varepsilon.$$

Now we minimalize σ_μ^2 with $\mu \in \mathfrak{M}_2$. In fact the expectation m of μ_3 is given by $m = n_p \beta_{p+1}(z_{p+1} - l)$, where

$$n_p = \begin{cases} p & \text{if } \ p \text{ is even,} \\ p+1 & \text{if } \ p \text{ is odd.} \end{cases}$$

Hence we get

$$\begin{aligned} \sigma_{\mu_3}^2 &= \sum_{2k=-p}^{p} (2kl)^2 \beta_p + \sum_{2k+1=-p}^{p} (2kl)^2 \beta_{p+1} - n_p^2 \beta_{p+1}^2 (z_{p+1} - l)^2 \\ &= \left(\sum_{k=0}^{m_p} k^2\right) 8l^2 \beta_p \\ &\quad + \beta_{p+1}\left[n_p z_{p+1}^2 + n_p^2 l^2 - 2n_p l z_{p+1} + 8l^2 \left(\sum_{k=0}^{(n_p/2)-1} k^2\right)\right] \\ &\quad - n_p^2 \beta_{p+1}^2 (z_{p+1} - l)^2, \end{aligned}$$

where

$$m_p = \begin{cases} p/2 & \text{if } \ p \text{ is even,} \\ (p-1)/2 & \text{if } \ p \text{ is odd.} \end{cases}$$

We have to find the minimum of the expression

$$\begin{aligned} &z_{p+1}^2 - 2l z_{p+1} - n_p \beta_{p+1}^2 z_{p+1}^2 + 2l z_{p+1} n_p \beta_{p+1} \\ &\qquad = z_{p+1}(z_{p+1} - 2l)(1 - n_p \beta_{p+1}); \end{aligned}$$

this minimum is reached for $z_{p+1} = l$.

Put $\mu_0 = \sum_{k=0}^{2p} \beta_k \, \delta_{(k+p)l}$. Then $\sigma_{\mu_0}^2 \leq \sigma_\mu^2 + \varepsilon$ for every $\varepsilon > 0$ and for every $\mu \in \mathfrak{M}_1$. Hence $\sigma_{\mu_0}^2 = \inf_{\mu \in \mathfrak{M}_1} \sigma_\mu^2$, and

$$\sigma_{\mu_0}^2 = (l^2/3)p(p+1)(3 - \alpha(2p+1)), \tag{1.6.2}$$

i.e., a linear function of α. ◇

Complement

It is easily seen that $\sigma_\mu \geq (1-p)^{1/2} Z_\mu(p)/2$ for any $p \in (0, 1)$.

Notes and Comments

The results of this section are due to Lévy (1937, 1954).

1.7 MULTIDIMENSIONAL CASE

1.7.1 Let us consider a p.m. μ on $\mathscr{B}^r$. We shall see that for $r > 1$ we can define several cn.f.'s associated with μ.

We begin by defining four real-valued functions $Q_\mu^{\square i}$, $1 \leq i \leq 4$, on R^r, by the same procedure as for $r = 1$ (cf. pp. 1–2). For instance,

$$Q_\mu^{\square 1}(\boldsymbol{l}) = \begin{cases} \sup\limits_{\boldsymbol{x} \in R^r} \mu((-\boldsymbol{l}/2, \boldsymbol{l}/2) + \boldsymbol{x}) & \text{if } \boldsymbol{l} \in R_+^r, \\ 0 & \text{otherwise,} \end{cases}$$

and

$$Q_\mu^{\square 4}(\boldsymbol{l}) = \begin{cases} \sup\limits_{\boldsymbol{x} \in R^r} \mu([-\boldsymbol{l}/2, \boldsymbol{l}/2] + \boldsymbol{x}) & \text{if } \boldsymbol{l} \in \operatorname{Cl} R_+^r, \\ 0 & \text{otherwise;} \end{cases}$$

here we have used the coordinatelike partial order in R^r, as well as a self-explanatory notation for " intervals " (rectangles) in R^r.

DEFINITION 1.7.1

The functions $Q_\mu^{\square i}$, $1 \leq i \leq 4$, are called *rectangular cn.f.'s.*

Clearly these functions are nondecreasing and bounded by 1. However, we have, as for $r = 1$, only two different cn.f.'s.

THEOREM 1.7.2

We have

(1) $Q_\mu^{\square 1} = Q_\mu^{\square 2} = Q_\mu^{\square 3} = Q_\mu^{\square -}$; moreover, $Q_\mu^{\square -}(\boldsymbol{l}) = 0$ for every $\boldsymbol{l} \notin R_+^r$, $Q_\mu^{\square -}(\infty) = 1$, and $Q_\mu^{\square -}$ is left continuous†;
(2) $Q_\mu^{\square 4} = Q_\mu^{\square}$ is right continuous;
(3) $Q_\mu^{\square}(\boldsymbol{0}) = \sup_{\boldsymbol{x} \in R^r} \mu(\{\boldsymbol{x}\})$;
(4) $Q_\mu^{\square}(\boldsymbol{0}) = 0$ iff μ is nonatomic;
(5) $Q_\mu^{\square} = Q_{\mu_a}^{\square} = Q_{\tilde{\mu}}^{\square}$, where $\mu_a(A) = \mu(A - a)$, and $\tilde{\mu}(A) = \mu(-A)$ for every $A \in \mathscr{B}^r$;
(6) for every $\boldsymbol{l} \geq \boldsymbol{0}$, there is $\boldsymbol{x_l}$ such that $Q_\mu^{\square}(\boldsymbol{l}) = \mu([-\boldsymbol{l}/2, \boldsymbol{l}/2] + \boldsymbol{x_l})$.

Proof

It suffices to make minor modifications in the proofs given in the case $r = 1$; for instance, replace the h's by $h\boldsymbol{e}$ in the proof of Proposition 1.1.3 in order to get (1) above. ◇

For $r = 1$, we proved (cf. Theorem 1.1.6) that $\mathfrak{Q} = \tilde{\mathfrak{F}}_+$. The following example shows that $Q_\mu^{\square}$, for $r > 1$, is in general not a d.f. Take $r = 2$, and

$$F(x_1, x_2) = \begin{cases} 0 & \text{if } x_1 < 0 \text{ or } x_2 < 0; \\ \frac{1}{4}(x_1 + x_2) & \text{if } 0 \leq x_1 < 1,\ 0 \leq x_2 < 1; \\ \frac{1}{4}x_1 + \frac{1}{4} & \text{if } 0 \leq x_1 < 1,\ x_2 \geq 1; \\ \frac{1}{4}x_2 + \frac{1}{4} & \text{if } x_1 \geq 1,\ 0 \leq x_2 < 1; \\ 1 & \text{if } x_1 \geq 1,\ x_2 \geq 1. \end{cases}$$

Then

$$Q_F^{\square}(l_1, l_2) = \begin{cases} 0 & \text{if } l_1 < 0 \text{ or } l_2 < 0; \\ \frac{1}{2} & \text{if } 0 \leq l_1 < 1,\ 0 \leq l_2 < 1; \\ \frac{1}{2}l_1 + \frac{1}{2} & \text{if } 0 \leq l_1 < 1,\ l_2 \geq 1; \\ \frac{1}{2}l_2 + \frac{1}{2} & \text{if } l_1 \geq 1,\ 0 \leq l_2 < 1; \\ 1 & \text{if } l_1 \geq 1,\ l_2 \geq 1. \end{cases}$$

† For a nondecreasing real-valued function Q on R^r, this means that $Q^-(\boldsymbol{l}) = \lim_{\delta \downarrow 0} Q(\boldsymbol{l} - \delta \boldsymbol{e}) = Q(\boldsymbol{l})$, where $\boldsymbol{e} = \underbrace{(1, \ldots, 1)}_{r \text{ times}}$; analogously we define right continuity. Q is continuous if it is left and right continuous.

Clearly, $Q_F^{\square}$ is not a d.f. since the second difference

$$Q_F^{\square}(1.2, 1.2) + Q_F^{\square}(0.8, 0.8) - Q_F^{\square}(1.2, 0.8) - Q_F^{\square}(0.8, 1.2) = -0.3.$$

However, $Q_\mu^{\square}$ is subadditive with respect to each of the arguments:

THEOREM 1.7.3

$$Q_\mu^{\square}(\boldsymbol{x} + l\boldsymbol{e}_k) \le Q_\mu^{\square}(\boldsymbol{x}) + Q_\mu^{\square}(\boldsymbol{x} + (l - x)_k \boldsymbol{e}_k), \tag{1.7.1}$$

for every $l \ge 0$ and $\boldsymbol{x} = (x_1, \ldots, x_r) \ge 0$, where $\boldsymbol{e}_k$ is the unit vector in the kth direction, $1 \le k \le r$.

Proof

By Theorem 1.7.2 (6), there is $\boldsymbol{y}_{l,k} \in R^r$ such that

$$\begin{aligned} Q_\mu^{\square}(\boldsymbol{x} + l\boldsymbol{e}_k) &= \mu([\boldsymbol{y}_{l,k}, \boldsymbol{y}_{l,k} + \boldsymbol{x} + l\boldsymbol{e}_k]) \\ &\le \mu([\boldsymbol{y}_{l,k}, \boldsymbol{y}_{l,k} + \boldsymbol{x}]) \\ &\quad + \mu([\boldsymbol{y}_{l,k} + x_k \boldsymbol{e}_k, \boldsymbol{y}_{l,k} + x_k \boldsymbol{e}_k + \boldsymbol{x} + (l - x_k)\boldsymbol{e}_k]) \\ &\le Q_\mu^{\square}(\boldsymbol{x}) + Q_\mu^{\square}(\boldsymbol{x} + (l - x_k)\boldsymbol{e}_k). \quad \diamond \end{aligned}$$

The above inequality is true for every $\boldsymbol{x} \ge \boldsymbol{0}$ and $l \ge 0$, and therefore it holds also for $Q_\mu^{\square -}$. We remark that $(\boldsymbol{x}, \boldsymbol{x} + \boldsymbol{l}) = \varnothing$ if $\boldsymbol{l} \in \partial R_+^r$.

COROLLARY

We have

$$Q_\mu^{\square}(\boldsymbol{l}_1 + \boldsymbol{l}_2) \le \sum_{k_1, \ldots, k_r = 1}^{2} Q_\mu^{\square}(l_{k_1, 1}, \ldots, l_{k_r, r}), \tag{1.7.2}$$

where $\boldsymbol{l}_k = (l_{k,1}, \ldots, l_{k,r})$, $k = 1, 2$.

Proof

In fact (1.7.2) is equivalent to (1.7.1). $\diamond$

We have seen (cf. Theorem 1.1.6) that for $r = 1$ and $F \in \tilde{\mathfrak{F}}_+$ we have $Q_F = F$. The following example shows that a d.f. such that $F^-(\boldsymbol{x}) = 0$ for $\boldsymbol{x} \notin R^r_+$, and which is subadditive with respect to each of the arguments, does not necessarily have this property. Indeed, take $r = 2$, and

$$\mu = \tfrac{1}{6}\,\delta_{(0,0)} + \tfrac{1}{6}\,\delta_{(1,0)} + \tfrac{1}{6}\,\delta_{(2,0)} + \tfrac{1}{12}\,\delta_{(0,1)} + \tfrac{1}{12}\,\delta_{(0,2)} + \tfrac{1}{3}\,\delta_{(10,10)}.$$

The corresponding d.f. F has the above properties; however, $Q_F^{\square} \neq F$, since $Q_\mu^{\square}(0, 0) = \frac{1}{3}$, and $F(0, 0) = \frac{1}{6} \neq \frac{1}{3}$.

Let us discuss now the problem of continuity. We have already seen [cf. Theorem 1.7.2 (4)] that $Q_\mu^{\square}(\boldsymbol{0}) = 0$ iff μ is nonatomic. For $r > 1$, this does not necessarily imply continuity of the d.f. associated with μ. However, it is possible to give a necessary and sufficient condition for continuity in terms of the cn.f. Indeed, we can prove

THEOREM 1.7.4

The d.f. associated with μ is continuous iff the restriction of $Q_\mu^{\square}$ to ∂R_r^+ vanishes.

Proof

Suppose that the d.f. F, associated with μ, is continuous, and let $\boldsymbol{l} \in \partial R^r_+$. Then there is $\boldsymbol{x_l} \in R^r$ such that

$$Q_\mu^{\square}(\boldsymbol{l}) = \mu([\boldsymbol{x_l}, \boldsymbol{x_l} + \boldsymbol{l}]) \leq F(\boldsymbol{x_l} + \boldsymbol{l}) - F^-(\boldsymbol{x_l} + \boldsymbol{l}) = 0;$$

hence the restriction of $Q_\mu^{\square}$ to ∂R^r_+ vanishes.

Conversely, suppose that F is not continuous. Then there is $\boldsymbol{x} = (x_1, \ldots, x_r) \in R^r$ such that $F(\boldsymbol{x}) - F^-(\boldsymbol{x}) = a > 0$. On the other hand, there is $M > 0$ such that $F(\boldsymbol{y}) < a/2$ if at least one $y_k < -M$. Take $l_k \geq x_k + M$, $1 \leq k \leq r$. Then there is an integer j, $1 \leq j \leq r$, such that $Q_\mu^{\square}\left(\sum_{k \neq j} l_k \boldsymbol{e}_k + \varepsilon \boldsymbol{e}_j\right) > a/2^r$ for every $\varepsilon > 0$ and therefore $Q_\mu^{\square}\left(\sum_{k \neq j} l_k \boldsymbol{e}_k\right) > a/2^r$, i.e., the restriction of $Q_\mu^{\square}$ to ∂R^r_+ does not vanish. ◇

Let us remark that since for monotonic functions continuity with respect to each argument implies global continuity, it is easily seen that $Q_\mu^{\square}$ is continuous iff the restriction of this function to ∂R^r vanishes.

An interesting function which may be obtained from $Q_\mu^{\square}$ is the restriction of this function to the set

$$\{\boldsymbol{l} = (l_1, \ldots, l_r) : l_k = l > 0, \quad 1 \leq k \leq r\},$$

which leads in fact to the *square cn.f.* $Q_\mu^{\square}$, i.e. a cn.f. defined by means of squares only, and which depends only on a single argument.

1.7.2 Further, let us set $\langle \boldsymbol{a}, \boldsymbol{b} \rangle = (-\infty, \boldsymbol{a}) - (-\infty, \boldsymbol{b}]$ for L-shaped intervals, $\boldsymbol{a} < \boldsymbol{b}$; we define four real-valued functions $Q_\mu^{\neg i}$, $1 \leq i \leq 4$, on R^r by putting, e.g., for $Q_\mu^{\neg 4}$:

$$Q_\mu^{\neg 4}(\boldsymbol{l}) = \begin{cases} \sup\limits_{\boldsymbol{x} \in R^r} \mu(\langle -\boldsymbol{l}/2, \boldsymbol{l}/2 \rangle + \boldsymbol{x}) & \text{if } \boldsymbol{l} \geq \boldsymbol{0}, \\ 0 & \text{otherwise.} \end{cases}$$

DEFINITION 1.7.5

The functions $Q_\mu^{\neg i}$, $1 \leq i \leq 4$, are called L-shaped cn.f.'s.

Again, we have $Q_\mu^{\neg 1} = Q_\mu^{\neg 2} = Q_\mu^{\neg 3} = Q_\mu^{\neg -}$ and $Q_\mu^{\neg 4} = Q_\mu^{\neg}$.

1.7.3 For many problems it is sufficient to consider cn.f.'s depending on a single argument. We have already seen that one example is given by $Q_\mu^{\square}$. But the square cn.f., e.g., is not the only possible choice. Indeed, let d be a distance induced by a norm on R^r and let us set

$$Q_\mu^{\bigcirc, d}(l) = \begin{cases} 0 & \text{if } l < 0, \\ \sup\limits_{\boldsymbol{x} \in R^r} \mu(\text{Cl } S(\boldsymbol{x}; l)) & \text{if } l \geq 0, \end{cases}$$

where $S(\boldsymbol{x}; l)$ is the open "sphere" with center $\boldsymbol{x}$ and radius $l/2$, i.e., $S(\boldsymbol{x}; l) = \{\boldsymbol{y} : d(\boldsymbol{x}, \boldsymbol{y}) < l/2\}$, and $\text{Cl } S(\boldsymbol{x}; l) = \{\boldsymbol{y} : d(\boldsymbol{x}, \boldsymbol{y}) \leq l/2\}$.

DEFINITION 1.7.6

The function $Q_\mu^{\bigcirc, d}$ is called a *d-spherical cn.f.*

Let us consider two special cases. First, let

$$d(\boldsymbol{x}, \boldsymbol{y}) = d_{\infty}(\boldsymbol{x}, \boldsymbol{y}) = \sup_{1 \le k \le r} |x_k - y_k|,$$

where $\boldsymbol{x} = (x_1, \ldots, x_r)$ and $\boldsymbol{y} = (y_1, \ldots, y_r)$; clearly $Q_\mu^{\square} = Q_\mu^{\bigcirc, d_\infty}$. Next, let

$$d(\boldsymbol{x}, \boldsymbol{y}) = d_2(\boldsymbol{x}, \boldsymbol{y}) = \left(\sum_{k=1}^{r} (x_i - y_i)^2 \right)^{1/2},$$

i.e., the Euclidean distance; then $S(\boldsymbol{x}; l)$ is an Euclidean sphere, and $Q_\mu^{\bigcirc} = Q_\mu^{\bigcirc, d_2}$ is called a *spherical cn.f.* Hence, $Q_\mu^{\square}$ and $Q_\mu^{\bigcirc}$ are in fact both d-spherical cn.f.'s, but with respect to different distances.

Obviously, for $Q_\mu^{\bigcirc}$, e.g., we can write a theorem analogous to Theorem 1.7.2, with the remark that in this case $Q_\mu^{\bigcirc}$ is a right continuous d.f.

We have also the obvious inequalities

$$Q_\mu^{\square}(\boldsymbol{l}) \le Q_\mu^{\bigcirc}(|\boldsymbol{l}|) \le Q_\mu^{\square}(|\boldsymbol{l}|, \ldots, |\boldsymbol{l}|),$$

$$Q_\mu^{\bigcirc}\left(\min_{1 \le k \le r} l_k \right) \le Q_\mu^{\square}(\boldsymbol{l}) \le Q_\mu^{\bigcirc}(|\boldsymbol{l}|),$$

and

$$Q_\mu^{\square}(\boldsymbol{l}) \le Q_\mu^{\urcorner}(\boldsymbol{l}), \qquad Q_\mu^{\bigcirc}\left(\min_{1 \le k \le r} l_k \right) \le Q^{\urcorner}(\boldsymbol{l}),$$

where $\boldsymbol{l} = (l_1, \ldots, l_r)$.

1.7.4 We can even associate with a p.m. μ on $\mathscr{B}^r$ a family of r cn.f.'s. Namely, let μ, $1 \le u \le r$, be given, and let $Q_{\mu, u}$ be a real-valued function defined on R by the relation

$$Q_{\mu, u}(l) = \begin{cases} 0 & \text{if } l < 0, \\ \sup_{A \in \mathscr{C}_u(l)} \mu(A) & \text{if } l \ge 0, \end{cases}$$

where $\mathscr{C}_u(l)$ is the class of all closed convex sets whose intersection with all possible u-dimensional hyperplanes is of u-dimensional volume not larger than l.

DEFINITION 1.7.7

The function $Q_{\mu,u}$ is called a *cn.f. of order u*, $1 \leq u \leq r$.

It is clear that for $r = 1$, all four definitions 1.7.1, 1.7.5–1.7.7, lead to the same notion, the Lévy cn.f.

Complements

1 Let μ be a p.m. on $\mathscr{B}^r$, and suppose that $\mu = \prod_{k=1}^{r} \mu_k$, where μ_k, $1 \leq k \leq r$, is a p.m. on $\mathscr{B}$. Then $Q_\mu^{\square} = \prod_{k=1}^{r} Q_{\mu_k}$ [Hengartner and Theodorescu (1972a)].

2 We have

$$Q_\mu^{\square}(\alpha_1 l_1, \ldots, \alpha_r l_r) \leq \left(\prod_{\alpha_k > 1} (2\alpha_k)\right) Q_\mu^{\square}(l_1, \ldots, l_r);$$

for $\alpha_k = \alpha > 1$, $1 \leq k \leq r$, we obviously get

$$Q_\mu^{\square}(\alpha \boldsymbol{l}) \leq 2^r \alpha^r Q_\mu^{\square}(\boldsymbol{l}).$$

Further, for $\alpha > 1$ and $l > 0$, we have

$$Q_\mu^{\bigcirc}(\alpha l) \leq (\text{const})\alpha^r Q_\mu^{\bigcirc}(l),$$

where const $\leq 2^r r^{r/2}$.

3 $Q_\mu^{\urcorner}$ is subadditive, and $Q_\mu^{\urcorner}(\boldsymbol{0}) = 0$ iff μ is continuous [Hengartner and Theodorescu (1972a)].

4 $Q_\mu(\boldsymbol{0}) = 0$ iff μ is nonatomic for any cn.f. defined by means of convex domains which reduce to a point as $\boldsymbol{l} \downarrow \boldsymbol{0}$.

5 Let $K(\boldsymbol{x};\ l,\ s) = \{z : s \leq |z - \boldsymbol{x}| \leq s + l/2\}$; clearly $\text{Cl}\, S(\boldsymbol{x};\ l) = K(\boldsymbol{x}; l, 0)$. Put

$$Q_\mu^{\circledcirc}(l) = \begin{cases} 0 & \text{if } \; l \leq 0, \\ \sup\limits_{s \in R^+} \; \sup\limits_{x \in R^r} \mu(K(x; l, s)) & \text{if } \; l \geq 0. \end{cases}$$

This function is subadditive, and may be also used as a cn.f.

Notes and Comments

Several possible definitions the cn.f. in the multidimensional case were suggested by different authors. Rectangular cn.f.'s are to be found in Esséen (1966), L-shaped cn.f.'s in Hengartner and Theodorescu (1972a), spherical cn.f.'s in Parthasarathy (1967) and Esséen (1968), and cn.f.'s of order u in Sebast'yanov (1963) and Sazonov (1966) [see also Prohorov and Rozanov (1969, p. 178)].

2 CONCENTRATION FUNCTIONS OF CONVOLUTIONS

This chapter examines the cn.f. of a convolution of p.m.'s, in other words, the cn.f. of a sum of independent r.v.'s. Section 2.1 is devoted to finding upper and lower bounds of the cn.f. of a convolution of p.m.'s in terms of the cn.f.'s of the factors. In Section 2.2 further upper bounds, known under the name *Kolmogorov type inequalities*, are given for the one-dimensional and for the multidimensional cases. Section 2.3 examines the special case of identical factors. Finally, Section 2.4 deals with asymptotic estimations when the number of factors is sufficiently large.

2.1 UPPER AND LOWER BOUNDS

2.1.1 Let $(\mu_n)_{n \in N^*}$ be a sequence of p.m.'s and set $\mu_{(n)} = \mathop{*}\limits_{k=1}^{n} \mu_k$, $n \in N^*$.

We begin by giving an upper bound for $Q_{\mu_{(n)}}$.

THEOREM 2.1.1
We have

$$Q_{\mu_{(n)}} \leq \min_{1 \leq k \leq n} Q_{\mu_k}. \tag{2.1.1}$$

Proof

We proceed by induction. Suppose $n = 2$; then

$$Q_{\mu_1 * \mu_2}(l) = \sup_{x \in R} \int_R \mu_1([-l/2, l/2] + x - z)\, d\mu_2(z) \leq Q_{\mu_1}(l),$$

and by commutativity of $*$,

$$Q_{\mu_1 * \mu_2} \leq Q_{\mu_2},$$

so that

$$Q_{\mu_1 * \mu_2} \leq \min(Q_{\mu_1}, Q_{\mu_2}).$$

Further, for $n > 2$, we have

$$\begin{aligned} Q_{\mu(n)} = Q_{\mu(n-1) * \mu_n} &\leq \min(Q_{\mu(n-1)}, Q_{\mu_n}) \\ &\leq \min_{1 \leq k \leq n} Q_{\mu_k}. \end{aligned}$$

In other words, the cn.f. can only decrease when passing from the factor p.m.'s to their convolution.

The above result holds also true for $Q_\mu^{\square}$ and Q_μ°.

Let us indicate now a lower bound for $Q_{\mu(n)}$.

THEOREM 2.1.2

We have

$$Q_{\mu(n)}(l) \geq \prod_{k=1}^{n} Q_{\mu_k}(\alpha_k l), \tag{2.1.2}$$

where $\alpha_k \geq 0$, $1 \leq k \leq n$, and $\sum_{k=1}^{n} \alpha_k = 1$.

Proof

Suppose first that $n = 2$. Then by (1.1.4) there exist $x_0, y_0 \in R$ such that

$$Q_{\mu_1}(\alpha_1 l) Q_{\mu_2}(\alpha_2 l) = \mu_1([x_0, x_0 + \alpha_1 l])\mu_2([y_0, y_0 + \alpha_2 l]).$$

On the other hand,

$$[x_0, x_0 + \alpha_1 l] \times [y_0, y_0 + \alpha_2 l] \\ \subset \{(x,y): x_0 + y_0 \leq x + y \leq x_0 + y_0 + l\};$$

therefore

$$Q_{\mu_1}(\alpha_1 l)Q_{\mu_2}(\alpha_2 l) \leq \mu_1 \times \mu_2(\{(x,y): x_0 + y_0 \leq x + y \leq x_0 + y_0 + l\}) \\ = \mu_1 * \mu_2([x_0 + y_0, x_0 + y_0 + l]) \leq Q_{\mu_1 * \mu_2}(l).$$

Suppose now $n > 2$. We proceed by induction, i.e.

$$\begin{aligned} Q_{\mu_{(n)}}(l) &\geq Q_{\mu_n}(\beta l)Q_{\mu_{(n-1)}}((1-\beta)l) \\ &\geq Q_{\mu_n}(\beta l)\prod_{k=1}^{n-1} Q_{\mu_k}((1-\beta)\alpha_k l) \\ &= \prod_{k=1}^{n} Q_{\mu_k}(\beta_k l); \end{aligned}$$

clearly $\beta_k \geq 0$, $1 \leq k \leq n$, and $\sum_{k=1}^{n} \beta_k = 1$. ◇

From Theorems 2.1.1 and 2.1.2 we deduce a result concerning continuity, namely

THEOREM 2.1.3

$\mu_{(n)}$ is nonatomic iff at least one of the μ_k is nonatomic.

Proof

Suppose that μ_{k_0} is nonatomic. Then by Theorems 2.1.1 and 1.2.2 we conclude that $Q_{\mu_{(n)}}(0) \leq \min_{1 \leq k \leq n} Q_{\mu_k}(0)$. Therefore $\mu_{(n)}$ is also nonatomic.

Conversely, suppose that $\mu_{(n)}$ is nonatomic. Then by (2.1.2) we have

$$\prod_{k=1}^{n} Q_{\mu_k}(l/n) \leq Q_{\mu_{(n)}}(l),$$

so that there exists a p.m., say μ_{k_0}, such that $Q_{\mu_{k_0}}(0) = 0$ and therefore, by Theorem 1.2.2 μ_{k_0} is nonatomic. ◇

2.1.2 The following result will be used further in characterizing the cn.f. of a convolution of p.m.'s.

PROPOSITION 2.1.4

We have

$$Q_\mu = H_0 \tag{2.1.3}$$

iff $\mu = \delta_a$.

Proof

Clearly, by Example (1), p. 12, $Q_\delta = H_0$. Suppose now (2.1.3) is valid. Then by (1.1.4) we get for all $l \geq 0$, $Q_\mu(l) = \mu([x_l, x_l + l]) = 1$. Choose a sequence $(l_n)_{n \in N^*}$ such that $l_n \downarrow 0$. Then there is a subsequence $(x_{l_n})_{n \in N^*}$ which converges to limit a with $\mu(\{a\}) = 1$, i.e., $\mu = \delta_a$. ◇

Now we can prove

THEOREM 2.1.5

We have

$$Q_{\mu * \nu} = Q_\mu \tag{2.1.4}$$

iff $\nu = \delta_a$.

Proof

If $\nu = \delta_a$, then the assertion follows from Theorem 1.4.1.

Conversely, suppose (2.1.4) true, and assume that $\nu \neq \delta_a$, that is there exists $x \in R$ such that $\nu((-\infty, x]) \in (0, 1)$.

Now, obviously $Q_\mu(l) \geq \mu([u - x, u - x + l])$ for all u, $x \in R$ and

$$\mu * \nu([u, u + l]) = \int_R \mu([u - x, u - x + l])\, d\nu(x) \leq Q_{\mu * \nu}(l)$$

for all $u \in R$. By (1.1.4) there exists $u_0 \in R$ such that $\mu * \nu([u_0, u_0 + l]) = Q_{\mu * \nu}(l) = Q_\mu(l)$, hence $\mu([u_0 - x, u_0 - x + l]) = Q_\mu(l)$ ν-a.s. Since $\nu \neq \delta_a$ there exist $x_1 < x_2$ such that $\mu([u_0 - x_i, u_0 - x_i + l]) = Q_\mu(l)$, $i = 1, 2$.

Let us set now $a = x_2 - x_1$; then we have

$$\begin{aligned} Q_\mu(l + a) &\geq \mu([u_0 - x_2, u_0 - x_1 + l]) \\ &\geq \mu([u_0 - x_1, u_0 - x_1 + l]) + \mu([u_0 - x_2, u_0 - x_2 + l]) - \varphi(l) \end{aligned}$$

where

$$\varphi(l) = \begin{cases} 0 & \text{if } l < a. \\ \mu([u_0 - x_1, u_0 - x_1 + l]) & \text{if } l \geq a. \end{cases}$$

Hence

$$Q_\mu(l + a) \geq \begin{cases} 2Q_\mu(l) & \text{if } l < a, \\ 2Q_\mu(l) - Q_\mu(l - a) & \text{if } l \geq a. \end{cases}$$

for all $l > 0$, and this independently of u_0. Consequently $Q_\mu(l + a) + Q_\mu(l - a) \geq 2Q_\mu(l)$ for all $l > 0$.

Next, take $l = an$, $n \in N^*$; then

$$Q_\mu((n + 1)a) \geq (n + 1)Q_\mu(a) - nQ_\mu(0).$$

If $Q_\mu(a) = 1$, then $Q_\mu(0) = 1$ since $2 = 2Q_\mu(a) \leq Q_\mu(2a) + Q_\mu(0) = 1 + Q_\mu(0)$, so that $Q_\mu = H_0$; by Proposition 2.1.4 it follows that $\mu = \delta_a$. Let $Q_\mu(a) < 1$; there is $l > 0$ such that $Q_\mu(l + a) - Q_\mu(l) = \eta > 0$. The sequence $Q_\mu(l + na) = \varphi_n(l)$, $n \in N^*$, is convex, i.e., $\varphi_{n+1} + \varphi_{n-1} \geq 2\varphi_n$, $n \in N^*$. Hence

$$\begin{aligned} Q_\mu(l + (n + 1)a) - Q_\mu(l + na) &\geq Q_\mu(l + na) - Q_\mu(l + (n - 1)a) \\ &\geq Q_\mu(l + a) - Q_\mu(l) = \eta > 0 \end{aligned}$$

and

$$1 \geq Q_\mu(l + (n + 1)a) \geq Q_\mu(l + na) + \eta \geq Q_\mu(l) + n\eta$$

for all $n \in N^*$, which leads to a contradiction. Thus $Q_\mu(l) = Q_{\mu * \nu}(l) = H_0(l) \leq Q_\nu(l)$, which contradicts that $\nu \neq \delta_a$. ◇

Let us give a simple consequence of this theorem:

COROLLARY

Let ξ and η be i.r.v.'s, $\xi \neq$ const a.s. Then ξ and $\xi + \eta$ are dependent.

Proof

Suppose ξ and $\xi + \eta$ independent. Then $\xi + \eta$ and $-\xi$ are independent, and therefore we get $Q_{\xi+\eta} \leq Q_\eta$ and $Q_\eta = Q_{(\xi+\eta)+(-\xi)} \leq Q_{\xi+\eta}$, i.e., $Q_{\xi+\eta} = Q_\eta$. Hence by Theorem 2.1.5. $\xi \equiv$ const a.s. ◇

2.1.3 What can we say if (2.1.4) holds true only for certain l? A partial answer is given by

THEOREM 2.1.6

Suppose that there exists an $l_0 > 0$ such that $Q_{\mu*\nu}(l_0) = Q_\mu(l_0)$ and set

$$a = \sup S_\nu - \inf S_\nu = x_2 - x_1,$$

where S_ν is the support of ν,

$$b = \sup S - \inf S = y_2 - y_1,$$

where

$$S = \{y : \mu([y, y + l_0]) = Q_\mu(l_0)\},$$

and

$$c = \sup T - \inf T = u_2 - u_1,$$

where

$$T = \{u : \mu * \nu([u, u + l_0]) = Q_{\mu*\nu}(l_0)\}.$$

Then $c \leq b - a$ provided that $a < \infty$.

Proof

We have

$$Q_\mu(l_0) = \mu * \nu([u_1, u_1 + l_0]) = \int_R \mu([u_1 - z, u_1 - z + l_0])\, d\nu(z),$$

$$Q_\mu(l_0) = \mu * \nu([u_2, u_2 + l_0]) = \int_R \mu([u_2 - z, u_2 - z + l_0])\, d\nu(z),$$

hence

$$\mu([u_1 - z, u_1 - z + l_0]) = Q_\mu(l_0), \qquad \nu\text{-a.s.},$$
$$\mu([u_1 - z + c, u_1 - z + c + l_0]) = Q_\mu(l_0), \qquad \bar{\nu}\text{-a.s.}$$

It follows that $y_1 \leq u_1 - x_2, y_2 \geq u_1 - x_1 + c$ so that $b = y_2 - y_1 \geq c + x_2 - x_1 = c + a$. ◇

COROLLARY

Suppose that there exists an $l_0 > 0$ such that $Q_{\mu * \nu(n)}(l_0) = Q_\mu(l_0)$ and that diam $S_{\nu_k} \geq a$, $1 \leq k \leq n$, where $a < \infty$. Let us set

$$b = \sup S - \inf S,$$

where

$$S = \{y : \mu([y, y + l_0]) = Q_\mu(l_0)]\}.$$

Then $b \geq na$.

Proof

We notice first that $S_{\nu(n)} \geq na$. Now by Theorem 2.1.6 we have $b \geq c + na \geq na$. ◇

Complements

1 We have the inequality

$$Q_{\mu * \nu} \leq Q_\mu Q_\nu + (1 - Q_\mu)(1 - Q_\nu)$$

due to Romanovsky [quoted after Linnik (1962, p. 8)].

2 Inequalities involving convolution, Kolmogorov–Smirnov distance, and modulus of continuity are to be found, e.g., in Lecam (1963, 1965a).

Notes and Comments

The property that the cn.f. can only decrease when passing to convolution (Theorem 2.1.1) is due to Lévy (1937, 1954) and is very important in dealing with convergence problems. Theorem 2.1.2 and Theorem 2.1.3 are due to Hengartner and Theodorescu. Proposition 2.1.4 as well as Theorems 2.1.5 and 2.1.6 are essentially due to Lévy (1937, 1954) [see also Sâmboan and Theodorescu (1968) for Proposition 2.1.4].

2.2 KOLMOGOROV TYPE INEQUALITIES

2.2.1 We begin by proving two auxiliary lemmas the first of which gives a lower and an upper bound for the cn.f. The second lemma will be applied to the upper bound in the first lemma.

LEMMA 2.2.1

For any positive l there are two positive absolute constants A_1 and A_2 such that

$$(2.2.1)\quad A_1\, l(1+bl)^{-1}\int_{|t|\leq b/2} |\psi_\mu(t)|^2 dt \leq Q_\mu(l) \leq A_2\, a^{-1}\int_{|t|\leq a} |\psi_\mu(t)|\, dt,$$

where b is an arbitrary positive parameter and a is a parameter satisfying $0 < al \leq 1$ but otherwise arbitrary; here ψ_μ is the c.f. of μ.

Proof

Let us introduce two auxiliary continuous real-valued functions u and U satisfying the following properties:

(a) $U(x) \geq 0$;

(b) $\int_R U(x)\, dx = 2\pi$;

(c) $u(t) = 0$ for $|t| \geq 1$;

(d) $u(t) = (2\pi)^{-1} \int_R e^{-itx}\, U(x)\, dx$.

Thus

$$|u(t)| \leq 1 \tag{2.2.2}$$

for every $t \in R$. Functions satisfying the above properties are for instance

$$U(x) = \left(\frac{\sin x/2}{x/2}\right)^2, \qquad u(t) = \begin{cases} 1 - |t| & \text{if} \quad |t| \leq 1, \\ 0 & \text{if} \quad |t| > 1. \end{cases} \tag{2.2.3}$$

We start with the easily proved relation

$$\int_R U(a(x - z))\, d\mu(x) = a^{-1} \int_{|t| \leq a} \psi_\mu(t) u(t/a) e^{-itz}\, dt, \tag{2.2.4}$$

where a, $z \in R$ and $a > 0$.

By (2.2.2) and (2.2.4) we get

$$\int_R U(a(x - z)\, d\mu(x) \leq a^{-1} \int_{|t| \leq a} |\psi_\mu(t)|\, dt. \tag{2.2.5}$$

Now let U be chosen such that

$$\min_{x \in [-l/2, l/2]} U(ax) \geq \rho > 0 \tag{2.2.6}$$

and set $I = [z - l/2, z + l/2]$. Then

$$\int_R U(a(x - z))\, d\mu(x) \geq \rho\mu(I)$$

or

$$\mu(I) \leq (a\rho)^{-1} \int_{|t| \leq a} |\psi_\mu(t)|\, dt.$$

Since z is arbitrary, we thus obtain

$$Q_\mu(l) \leq (a\rho)^{-1} \int_{|t| \leq a} |\psi_\mu(t)|\, dt.$$

From now on we choose U according to (2.2.3). Since $\sin x/x \geq 4 \sin \frac{1}{4}$ for $|x| \leq \frac{1}{4}$, when $0 < al \leq 1$, (2.2.6) is satisfied with $\rho = (4 \sin \frac{1}{4})^2 \approx (0.9896)^2 \geq 0.978$. Consequently

$$Q_\mu(l) \leq A_2 a^{-1} \int_{|t| \leq a} |\psi_\mu(t)| \, dt \tag{2.2.7}$$

where $A_2 = (4 \sin \frac{1}{4})^{-2} \approx 1.021$ is an absolute constant.

The left-hand side of the inequality (2.2.1) may also be proved by means of the relation (2.2.4) if we replace μ by $\mu^s = \mu * \tilde{\mu}$, thus ψ_μ by $|\psi_\mu|^2$, and set $z = 0$. The following method, however, is simpler.

Let the p.m. λ have the c.f. $u(t/b)$ and consider the p.m. $\nu = \mu^s * \lambda$ with the c f. $|\psi_\mu(t)|^2 u(t/b)$. Then

$$Q_\mu(l) \geq Q_\nu(l). \tag{2.2.8}$$

From the inversion formula for c.f.'s we get

$$\nu([-b^{-1}, b^{-1}]) = (\pi b)^{-1} \int_{|t| \leq b} |\psi_\mu(t)|^2 (1 - |t|/b) \frac{\sin t/b}{t/b} \, dt.$$

Since $\sin x/x \geq 2\pi^{-1}$ for $|x| \leq \pi/2$ we have

$$Q_\nu(2b^{-1}) \geq \nu([-b^{-1}, b^{-1}]) \geq b^{-1}\pi^{-2} \int_{|t| \leq b/2} |\psi_\mu(t)|^2 \, dt. \tag{2.2.9}$$

But we have for all $l > 0$

$$\begin{aligned} Q_\nu(2b^{-1}) &\leq (2(bl)^{-1} + 1) Q_\nu(l) \\ &\leq 2(1 + bl)(bl)^{-1} Q_\nu(l), \end{aligned}$$

and hence from (2.2.8) and (2.2.9)

$$Q_\mu(l) \geq (2\pi^2)^{-1} l(1 + bl)^{-1} \int_{|t| \leq b/2} |\psi_\mu(t)|^2 \, dt$$

which leads to the desired inequality. ◇

The factor $l(1 + bl)^{-1}$ of the left-hand side of (2.2.1) tends to zero as $l \to 0$; this must necessarily be the case, for instance, for nonatomic p.m.'s. On the other hand,

$$\lim_{l\to\infty}(2\pi^2)^{-1}l(1+bl)^{-1}\int_{|t|\le b/2}|\psi_\mu(t)|^2dt \le (2\pi^2)^{-1}<1$$

that is compatible with $Q_\mu(+\infty)=1$.

The following lemma is stated in such a way that it can be immediately applied to an infinitely divisible p.m. The measures M_k, $1\le k\le n$, of the lemma have the same properties as the measure occurring in Lévy's canonical representation of the c.f. of an infinitely divisible p.m.†

LEMMA 2.2.2

Let M_k, $1\le k\le n$, be nonnegative measures on the Borel sets of $R-\{0\}$ with the following properties:

(*i*) $M_k([-\varepsilon,\varepsilon]^c)<\infty$, for every $\varepsilon>0$;

(*ii*) $\int_{|x|\le 1} x^2\,dM_k(x)<\infty$, $1\le k\le n$.

Further, let $\kappa>0$, and let $\sigma_k^2\ge 0$, $1\le k\le n$, be constants. For any positive $l_1,\dots,l_n\le L$ we have

(2.2.10)

$$\int_{|t|\le 1/L}\exp\left\{-\kappa\sum_{k=1}^{n}\left(\tfrac{1}{2}\sigma_k^2t^2+\int_R(1-\cos tx)\,dM_k(x)\right)\right\}dt$$
$$\le A\kappa^{-1/2}\left\{\sum_{k=1}^{n}\left(\sigma_k^2+\int_{|x|<l_k}x^2\,dM_k(x)+l_k^2\int_{|x|\ge l_k}dM_k(x)\right)\right\}^{-1/2}$$

where A is a positive absolute constant.

† Let μ be a p.m. Then μ is infinitely divisible iff its c.f. ψ_μ can be written in the form

$$\psi_\mu(t)=\exp\left\{iat-\tfrac{1}{2}\sigma^2t^2+\int_R\left(e^{itx}-1-itx/(1+x^2)\right)dM(x)\right\}$$

where a and σ are constants, and M is a nonnegative measure on the Borel sets of $R-\{0\}$ with the properties:

(i) $M([-\varepsilon,\varepsilon]^c)<\infty$ for every $\varepsilon>0$;

(ii) $\int_{|x|\le 1}x^2\,dM(x)<\infty$.

This representation is unique.

Proof

Applying the inequality $1 - \cos x \geq \frac{11}{24} x^2$ for $|x| \leq 1$ we have for $|t| \leq 1/L$ and $1 \leq k \leq n$.

$$\text{(2.2.11)} \quad \begin{aligned} \int_R (1 - \cos tx)\, dM_k(x) &= \int_{|x|<l_k} (1 - \cos tx)\, dM_k(x) \\ &\quad + \int_{|x|\geq l_k} (1 - \cos tx)\, dM_k(x) \\ &\geq \tfrac{11}{24} t^2 \int_{|x|<l_k} x^2\, dM_k(x) \\ &\quad + \int_{|x|\geq l_k} (1 - \cos tx)\, dM_k(x). \end{aligned}$$

Let us denote by I the left-hand side of (2.2.10) and let us introduce the quantities

$$\text{(2.2.12)} \qquad \gamma_k = \int_{|x|<l_k} x^2\, dM_k(x), \qquad \sigma^2 = \sum_{k=1}^{n} (\sigma_k^2 + \gamma_k),$$

$$p_k = \int_{|x|\geq l_k} dM_k(x), \qquad 1 \leq k \leq n.$$

From (2.2.11) we obtain

(2.2.13)

$$I \leq \int_{|t|\leq 1/L} \exp\{-\tfrac{11}{24} \kappa\sigma^2 t^2\} \prod_{k=1}^{n} \exp\left\{-\kappa \int_{|x|\geq l_k} (1 - \cos tx)\, dM_k(x)\right\} dt.$$

Let

$$\alpha_0 = \kappa\sigma^2, \qquad \alpha_k = \kappa l_k^2 p_k, \qquad 1 \leq k \leq n,$$

(2.2.14)

$$\gamma = \sum_{k=0}^{n} \alpha_k = \kappa \sum_{k=1}^{n} \left(\sigma_k^2 + \int_{|x|<l_k} x^2\, dM_k(x) + l_k^2 \int_{|x|\geq l_k} dM_k(x)\right),$$

$$\beta = \alpha_k/\gamma, \qquad 0 \leq k \leq n.$$

As is easily seen, we may suppose that all β_k's are positive. From Hölder's inequality applied to (2.2.13) we get

$$\text{(2.2.15)} \quad I \leq \left(\int_{|t|\leq 1/L} \exp\{-\tfrac{11}{24} \gamma t^2\}\, dt\right)^{\beta_0}$$

$$\times \prod_{k=1}^{n} \left(\int_{|t|\leq 1/L} \exp\left\{-\gamma l_k^{-2} \int_R (1 - \cos tx)\, dN_k(x)\right\} dt\right)^{\beta_k},$$

where

$$dN_k(x) = \begin{cases} 0 & \text{if } |x| < l_k, \\ dM_k(x)/p_k & \text{if } |x| \geq l_k, \end{cases}$$

$1 \leq k \leq n$, is a p.m.

We shall estimate each of the integrals of the right-hand side of (2.2.15). We easily find that

$$I_0 = \int_{|t| \leq 1/L} \exp\{-\tfrac{11}{24}\gamma t^2\}\, dt \leq A\gamma^{-1/2}. \tag{2.2.16}$$

To estimate

$$I_k = \int_{|t| \leq 1/L} \exp\left\{-\gamma l_k^{-2} \int_R (1 - \cos tx)\, dN_k(x)\right\} dt, \qquad 1 \leq k \leq n,$$

we apply Jensen's inequality† for continuous convex functions. We obtain

$$\exp\left\{-\gamma l_k^{-2} \int_R (1 - \cos tx)\, dN_k(x)\right\} \leq \int_R \exp\{-\gamma l_k^{-2}(1 - \cos tx)\}\, dN_k(x)$$
$$1 \leq k \leq n.$$

Thus, changing the order of integration we get

$$I_k \leq \int_{|x| \geq l_k} \left(\int_{|t| \leq 1/L} \exp\{-\gamma l_k^{-2}(1 - \cos tx)\}\, dt \right) dN_k(x), \qquad 1 \leq k \leq n.$$

For $|x| \geq l_k$ we get

$$\begin{aligned}
&\int_{|t| \leq 1/L} \exp\{-\gamma l_k^{-2}(1 - \cos tx)\}\, dt \\
&\quad = |x|^{-1} \int_{|u| \leq |x|/L} \exp\{-\gamma l_k^{-2}(1 - \cos u)\}\, du \\
&\quad \leq 2|x|^{-1}(\text{int}[|x|/L\pi] + 1) \int_0^{\pi} \exp\{-\gamma l_k^{-2}(1 - \cos u)\}\, du \\
&\quad \leq (4/L\pi) \int_0^{\pi} \exp\{-\gamma l_k^{-2} u^2/6\}\, du \leq (4/L\pi) \int_R \exp\{-\gamma l_k^{-2} u^2/6\}\, du \\
&\quad \leq A\gamma^{-1/2}, \qquad 1 \leq k \leq n,
\end{aligned}$$

† Let μ be a p.m., let g be a positive and convex function, and let ϕ a μ^s-integrable real-valued function of a r.v. ξ. Then $g(E\phi(\xi)) \leq Eg(\phi(\xi))$.

and hence

$$I_k \leq A\gamma^{-1/2}, \qquad 1 \leq k \leq n. \tag{2.2.17}$$

From (2.2.15), (2.2.16), and (2.2.17) we finally obtain $I \leq A\gamma^{-1/2}$, γ being defined by (2.2.14), and the lemma is proved.† ◇

Before using Lemmas 2.2.1 and 2.2.2 in order to get our basic inequality (2.2.23), let us indicate here as on application an inequality for the cn.f. of an infinitely divisible p.m.

THEOREM 2.2.3

Let μ be an infinitely divisible p.m., and let the Levy canonical representation of its c.f. be

$$\psi_\mu(t) = \exp\{iat - \tfrac{1}{2}\sigma^2 t^2\} + \int_R (e^{itx} - 1 - itx/(1 - x^2))\, dM(x). \tag{2.2.18}$$

There are two positive absolute constants A_3 and A_4 such that

$$\begin{aligned} &A_3 \min\left(1, l\left(\sigma^2 + \int_{|x|<l} x^2\, dM(x)\right)^{-1/2}\right) \exp\left\{-\int_{|x|\geq l} dM(x)\right\} \\ &\leq Q_\mu(l) \leq A_4\, l\left(\sigma^2 + \int_{|x|<l} x^2\, dM(x) + l^2 \int_{|x|\geq l} dM(x)\right)^{-1/2} \end{aligned} \tag{2.2.19}$$

Proof

We begin by proving the right-hand side inequality, Since

$$|\psi_\mu(t)| = \exp\left\{-\tfrac{1}{2}\sigma^2 t^2 - \int_R (1 - \cos tx)\, dM(x)\right\},$$

it follows from Lemma 2.2.1 with $a = l^{-1}$ that

$$Q_\mu(l) \leq al \int_{|x|\leq 1/l} \exp\left\{-\tfrac{1}{2}\sigma^2 t^2 - \int_R (1 - \cos tx)\, dM(x)\right\} dt$$

and the desired inequality immediately results from Lemma 2.2.2 with $n = 1$.

† In the above proof the A's may be different. However, they are positive absolute constants.

To prove the left-hand side inequality we write ψ_μ as

$$\psi_\mu(t) = \exp\left\{ia't - \tfrac{1}{2}\sigma^2 t^2 + \int_{|x|<l} (e^{itx} - 1 - itx/(1+x^2))\, dM(x)\right\}$$
$$\times \exp\left\{\int_{|x|\geq l} (e^{itx} - 1)\, dM(x)\right\} = \psi_{\mu_1}(t)\psi_{\mu_2}(t),$$

where a' is a new constant. Thus $\mu = \mu_1 * \mu_2$. Let now

$$c = \int_{|x|\geq l} dM(x) \text{ and } dG(x) = dM(x)/c \text{ if } |x| \geq l \text{ and } dG(x) = 0$$

if $|x| < l$. Then

$$\psi_{\mu_2} = \int_R \exp(itx) \sum_{k=0}^{\infty} (c^k/k!)e^{-c}\, dG_{(n)}(x),$$

where $G_{(n)} = \mathop{\ast}\limits_{k=1}^{n} G_k$, $G_k = G$, $1 \leq k \leq n$, $G_{(0)} = H_0$. Consequently, we obtain

$$\mu_2(\{0\}) \geq \exp\left\{-\int_{|x|\geq l} dM(x)\right\}. \tag{2.2.20}$$

But by Theorem 2.1.2

$$Q_\mu(l) = Q_{\mu_1 * \mu_2}(l) \geq Q_{\mu_1}(l)\mu_2(\{0\}). \tag{2.2.21}$$

A lower bound for Q_{μ_1} is now obtained from Lemma 2.2.1. We find with $b = l^{-1}$ that

(2.2.22)

$$Q_{\mu_1}(l) \geq \tfrac{1}{2} A_1 l \int_{|x|\leq 1/2l} \exp\left\{-\sigma^2 t^2 - 2\int_{|x|<l} (1 - \cos tx)\, dM(x)\right\} dt$$
$$\geq \tfrac{1}{2} A_1 l \int_{|x|\leq 1/2l} \exp\left\{-\sigma^2 t^2 - t^2 \int_{|x|<l} x^2\, dM(x)\right\} dt.$$

From (2.2.20), (2.2.21), and (2.2.22) the left-hand side of (2.2.19) is easily obtained. ◇

Note that if we apply Lemma 2.2.1 directly to ψ_μ in order to find a lower bound we get the factor

$$\exp\left\{-4\int_{|x|\geq l} dM(x)\right\}$$

instead of

$$\exp\left\{-\int_{|x|\geq l} dM(x)\right\}.$$

COROLLARY

The infinitely divisible p.m. μ with the c.f. (2.2.18) has at least one mass point iff $\sigma = 0$ and $\int_R dM(x) < \infty$.

Proof

It follows directly from (2.2.19).

2.2.2 Let $(\mu_n)_{n\in N^*}$ be a sequence of p.m.'s and let us set, as before, $\mu_{(n)} = \mathop{*}\limits_{k=1}^{n} \mu_k$, $n \in N^*$. Moreover, for the sake of simplicity, let us set also $\mu_n^s = \mu_n * \tilde{\mu}_n$, $n \in N^*$. We shall combine Lemmas 2.2.1 and 2.2.2 in order to obtain a basic inequality for the cn.f. of $\mu_{(n)}$, $n \in N^*$. To do this, we need still an auxiliary notion.

By the *censored variance* (at $l \geq 0$) of a p.m. μ we understand the quantity $D^2(\mu, l)$ defined by

$$D^2(\mu; l) = \begin{cases} \mu(\{0\})^c) & \text{if } \ l = 0, \\ l^{-2}\int_{|x|<l} x^2\, d\mu(x) + \int_{|x|\geq l} d\mu(x) & \text{if } \ 1 > 0. \end{cases}$$

It is not difficult to show that

(i) $D^2(\mu; l)$ is a nonincreasing function of l;

(ii) $D^2(\mu; l) = 0$ for some $l \geq 0$ iff $\mu = \delta_0$;

(iii) $D^2(\mu; l) \geq u^{-2}\int_{|x|<u} x^2\, d\mu(x)$ for $u \geq l$.

Now we can prove our main result.

THEOREM 2.2.4

For any positive $l_1, \ldots, l_n \leq L$ we have

$$Q_{\mu_{(n)}}(L) \leq AL\left(\sum_{k=1}^{n} l_k^2\, D^2(\mu_k^s; l_k)\right)^{-1/2}, \tag{2.2.23}$$

where A is a positive absolute constant.

Proof

Clearly $\mu_{(n)}$ has the c.f. $\prod_{k=1}^{n} \psi_{\mu_k}$. From the right-hand side of Lemma 2.2.1 applied to $\mu_{(n)}$ we get with $a = L^{-1}$

$$Q_{\mu_{(n)}}(L) \leq A_2 L \int_{|t| \leq 1/L} \prod_{k=1}^{n} |\psi_{\mu_k}(t)|\, dt. \tag{2.2.24}$$

Using the inequality

$$|\psi_{\mu_k}(t)| \leq \exp\{-\tfrac{1}{2}(1 - |\psi_{\mu_k}(t)|^2)\}$$

and observing that

$$1 - |\psi_{\mu_k}(t)|^2 = \int_R (1 - \cos tx)\, d\mu_k^s(x),$$

we obtain from (2.2.24)

$$Q_{\mu_{(n)}}(L) \leq A_2 L \int_{|t| \leq 1/L} \exp\left\{\tfrac{1}{2} \sum_{k=1}^{n} \int_R (1 - \cos tx)\, d\mu_k^s(x)\right\} dt \tag{2.2.25}$$

The integral above is of the type encountered in Lemma 2.2.2 with the exception that μ_k^s may have a mass point at the origin. It is, however, easily seen that Lemma 2.2.2 is still applicable. Using (2.2.10) we get

$$Q_{\mu_{(n)}}(L) \leq AL\left(\sum_{k=1}^{n}\left(\int_{|x| < l_k} x^2\, d\mu_k^s(x) + l_k^2 \int_{|x| \geq l_k} d\mu_k^s(x)\right)\right)^{-1/2}$$

and the theorem is proved. ◇

For use in later considerations we shall give another inequality for $Q_{\mu_{(n)}}$. We start with

LEMMA 2.2.5

For any positive l, we have

$$Q_{\mu_{(n)}}(l) \le Aa^{-1} \int_{|t| \le a} \exp\left\{ -\tfrac{11}{48} t^2 \sum_{k=1}^{n} \chi_k(t) \right\} dt,$$

where $0 < al \le 1$ and

$$\chi_k(t) = \int_{|x| \le 1/|t|} x^2 \, d\mu_k^s(x), \qquad 1 \le k \le n. \tag{2.2.27}$$

Proof

Put $L = a^{-1}$ in (2.2.25). Then we get for all $0 < l \le L$

$$Q_{\mu_{(n)}}(l) \le A_2 a^{-1} \int_{|t| \le a} \exp\left\{ -\tfrac{1}{2} \sum_{k=1}^{n} \int_R (1 - \cos tx) \, d\mu_k^s(x) \right\} dt.$$

To the right-hand side of the above inequality we apply the inequalities

$$\int_R (1 - \cos tx) \, d\mu_k^s(x) \ge \int_{|x| \le 1/|t|} (1 - \cos tx) \, d\mu_k^s(x) \ge \tfrac{11}{24} t^2 \chi_k(t),$$
$$1 \le k \le n,$$

so that we have proved the above lemma. ◇

THEOREM 2.2.6

For any positive $l \le L$ we have

$$Q_{\mu_{(n)}}(L) \le A(L/l) \left(\sup_{u \ge l} u^{-2} \sum_{k=1}^{n} \int_{|x| \le u} x^2 \, d\mu_k^s(x) \right)^{-1/2}, \tag{2.2.28}$$

where A is a positive absolute constant.

Proof

By Theorem 1.5.3, it is sufficient to prove (2.2.28) for $l = L$. Since χ_k is an even function and nonincreasing for $t > 0$, we get from (2.2.26)

$$Q_{\mu_{(n)}}(l) \le Aa^{-1} \int_{|t| \le a} \exp\left\{ -\tfrac{11}{48} t^2 \sum_{k=1}^{n} \chi_k(a) \right\} dt \le Aa^{-1} \left(\sum_{k=1}^{n} \chi_k(a) \right)^{-1/2},$$

where $0 < al \leq 1$. Putting $a = u^{-1}$ we obtain from the last inequality and the definition (2.2.27) of χ_k the desired result. ◇

We have evidently the following

COROLLARY

For any positive $l \leq L$ and for $\mu_1 = \cdots = \mu_\mu = \mu$ with μ nondegenerate we have

$$Q_{\mu_{(n)}}(L) \leq A(L/l)\left(\sup_{u \geq l} u^{-2} \int_{|x| \leq u} x^2 \, d\mu^s(x)\right)^{-1/2} n^{-1/2},$$

where A is a positive absolute constant.

Theorem 2.2.6 follows from Theorem 2.2.4, and property (iii) of the censored variance stated at the beginning of this section. We chose, however, to prove the inequality (2.2.28) directly since the proof generalizes to the multidimensional case.

2.2.3 Let us mention several special cases of Theorem 2.2.4.

PROPOSITION 2.2.7

For any positive $l_1, \ldots, l_n \leq L$ we have

$$Q_{\mu_{(n)}}(L) \leq AL\left(\sum_{k=1}^{n} l_k^2 \, \mu_k^s((-l_k/2, l_k/2,)^c)\right)^{-1/2}, \tag{2.2.29}$$

where A is a positive absolute constant.

Proof

It suffices to remark that $D^2(\mu; l) \geq \mu((-l/2, l/2)^c)$. ◇

PROPOSITION 2.2.8

For any positive numbers $l_1, \ldots, l_n \leq L$ assume that there are b_k and $a_k > 0$, $1 \leq k \leq n$, such that

$$\mu_k((-\infty, b_k - l/2]) \geq a_k, \qquad \mu_k([b_k + l/2, \infty)) \geq a_k, \qquad 1 \leq k \leq n.$$

Then we have

$$Q_{\mu(n)}(L) \leq AL\left(\sum_{k=1}^{n} l_k^2 a_k\right)^{-1/2}, \tag{2.2.30}$$

where A is a positive absolute constant.

Proof

Let us first remark that if μ is a p.m. such that

$$\mu((-\infty, b - l/2]) \geq a, \qquad \mu([b + l/2, \infty)) \geq a, \qquad l \geq 0,$$

then

$$\mu^s((-l/2, l/2)^c) \geq a/2. \tag{2.2.31}$$

Indeed, we can assume that $b = 0$. Then

$$\mu^s((-l/2, l/2)^c) \geq \mu([l/2, \infty))\tilde{\mu}([0, \infty)) + \mu((-\infty, l/2))\tilde{\mu}((-\infty, 0]).$$

If med $\mu \geq 0$, we get

$$\mu^s((-l/2, l/2)^c) \geq a/2.$$

The case med $\mu \leq 0$ is treated similarly. Consequently, (2.2.30) follows from (2.2.29) if we take into account (2.2.31). ◇

The following inequality is known as the Kolmogorov–Rogozin inequality.

PROPOSITION 2.2.9

For any positive $l_1, \ldots, l_n \leq L$ we have

$$Q_{\mu(n)}(L) \leq AL\left(\sum_{k=1}^{n} l_k^2 (1 - Q_{\mu_k}(l_k))\right)^{-1/2}, \tag{2.2.32}$$

where A is a positive absolute constant.

Proof

From

$$\mu^{s}((-l/2, l/2)^{c}) \geq 1 - Q_{\mu^{s}}(l) \geq 1 - Q_{\mu}(l)$$

it is seen that (2.2.29) implies (2.2.32). ◇

PROPOSITION 2.2.10

For any positive $l \leq L$, we have

$$Q_{\mu_{(n)}}(L) \leq A(L/l)\left(\sum_{k=1}^{n} (1 - Q_{\mu_k}(l))\right)^{-1/2}, \tag{2.2.33}$$

where A is a positive absolute constant.

Proof

Take in (2.2.30) $l_1 = \cdots = l_n = l$. ◇

PROPOSITION 2.2.11

For any positive $l \leq L$ we have

$$Q_{\mu_{(n)}}(L) \leq A(L/l)\left(\sum_{k=1}^{n} D^2(\mu_k^{s}; l)\right)^{-1/2}, \tag{2.2.34}$$

where A is a positive absolute constant. If, furthermore, $l = L$ and $\mu_1 = \cdots = \mu_n = \mu$ with μ nondegenerate, then

$$Q_{\mu_{(n)}}(l) \leq A(\mu; l)n^{-1/2}, \tag{2.2.35}$$

where $A(\mu; l)$ is a positive constant independent of n but depending on μ and on l.

Proof

Take first $l_1 = \cdots = l_n = l$ in (2.2.23) and then $\mu_1 = \cdots = \mu_n = \mu$; here $A(\mu; l) = A(D(\mu^{s}; l))^{-1}$. ◇

Let us remark that (2.2.35) follows also from the corollary to Theorem 2.2.6.

The following inequality is known as the Doeblin–Lévy inequality.

PROPOSITION 2.2.12

For every $0 < \varepsilon \leq 1$ and $\eta > 0$ there exist $h > 0$ and $n_0 \in N^*$ such that for every $n \geq n_0$ the inequalities

$$Q_{\mu_k}(l_k) \leq 1 - \varepsilon \tag{2.2.36}$$

for every $k \in N^*$ imply

$$Q_{\mu_{(n)}}\left(h\left(\sum_{k=1}^{n} l_k^2\right)^{1/2}\right) \leq \eta, \tag{2.2.37}$$

provided that $0 < l_k \leq M$, $k \in N^*$. Moreover, if $\eta \geq 1 - \varepsilon$, then there exists $h_0 > 0$ such that (2.2.37) is valid with $h = h_0$ for every $n \in N^*$.

Proof

We distinguish two cases:

$$\text{(a)} \quad \sum_{k \in N^*} l_k^2 = \infty, \qquad \text{and} \qquad \text{(b)} \quad \sum_{k \in N^*} l_k^2 < \infty.$$

(a) From (2.2.32) and (2.2.36)

$$Q_{\mu_{(n)}}(L) \leq AL\left(\varepsilon \sum_{k=1}^{n} l_k^2\right)^{-1/2}.$$

Now let $L = h\left(\sum_{k=1}^{n} l_k^2\right)^{1/2}$; we obtain

$$Q_{\mu_{(n)}}\left(h\left(\sum_{k=1}^{n} l_k^2\right)^{1/2}\right) \leq Ah\varepsilon^{-1/2}$$

ε and η being given, we determine h such that $Ah\varepsilon^{-1/2} \leq \eta$, i.e., $h \leq \eta\varepsilon^{1/2}A^{-1}$, and this value does not depend on $n \in N^*$. But (2.2.32) was proved when

$$\max_{1 \leq k \leq n} l_k \leq h\left(\sum_{k=1}^{n} l_k^2\right)^{1/2}.$$

Since $\sum_{k \in N^*} l_k^2 = \infty$, there exists $n_0 \in N^*$ such that

$$\max_{1 \le k \le n} l_k \le h\left(\sum_{k=1}^{n} l_k^2\right)^{1/2}$$

for every $n \ge n_0$.

(b) If $\sum_{k \in N^*} l_k^2 < \infty$, there exists $n_0 \in N^*$ such that $\sum_{k \in N^*} l_k^2 \le n l_1^2$ for every $n \ge n_0$. Take now $l_k' = l_1$, $k \in N^*$. Then there exists $n_1 \ge n_0$ such that for every $n \ge n_1$ and for the same h as in (a), we have by (2.2.33)

$$Q_{\mu(n)}\left(h\left(\sum_{k=1}^{n} l_k^2\right)^{1/2}\right) \le Q_{\mu(n)}(hn^{1/2} l_1) \le \eta, \qquad hn_1^{1/2} > 1.$$

Suppose now $\eta \ge 1 - \varepsilon$. Since $Q_{\mu_1}(l_1) \le 1 - \varepsilon \le \eta$, it follows by Theorem 2.1.1 that $Q_{\mu(n)}(l_1) \le \eta$ for every $n \in N^*$. From the first part of this proposition we know that there exist $h > 0$ and $n_0 \in N^*$ such that

$$Q_{\mu(n)}\left(h\left(\sum_{k=1}^{n} l_k^2\right)^{1/2}\right) \le \eta$$

for every $n \ge n_0$. Let us set now

$$h_0 = \min\left(h, l_1\left(\sum_{k=1}^{n_0} l_k^2\right)^{-1/2}\right).$$

Then

$$Q_{\mu(n)}\left(h_0\left(\sum_{k=1}^{n} l_k^2\right)^{1/2}\right) \le Q_{\mu(n)}\left(h\left(\sum_{k=1}^{n} l_k^2\right)^{1/2}\right) \le \eta$$

if $n \ge n_0$, and

$$Q_{\mu(n)}\left(h_0\left(\sum_{k=1}^{n} l_k^2\right)^{1/2}\right) \le Q_{\mu(n)}\left(l_1\left(\sum_{k=1}^{n} l_k^2\right)^{1/2}\left(\sum_{k=1}^{n_0} l_k^2\right)^{-1/2}\right)$$
$$\le Q_{\mu(n)}(l_1) \le \eta$$

if $n \le n_0$. ◇

PROPOSITION 2.2.13

For every $0 < \varepsilon \le 1$ and $\eta > 0$ there exist $h > 0$ and $n_0 \in N^*$ such that for every $n \ge n_0$ the inequalities

$$Q_{\mu_k}(l) \le 1 - \varepsilon$$

for every $k \in N^*$ imply

$$Q_{\mu_{(n)}}(hln^{1/2}) \leq \eta, \tag{2.2.38}$$

provided that $0 < l < \infty$. Moreover, if $\eta \geq 1 - \varepsilon$, then there exists $h_0 > 0$ such that (2.2.38) is valid with $h = h_0$ for every $n \in N^*$.

Proof

Take $l_k = l$, $k \in N^*$ in Proposition 2.2.12. ◇

PROPOSITION 2.2.14

Let μ_k, $1 \leq k \leq n$, be atomic p.m.'s. Then there is a positive absolute constant A such that

$$\sup_{x \in R} \mu_{(n)}(\{x\}) \leq A\left(\sum_{k=1}^{n}\left(1 - \sup_{x \in R} \mu_k(\{x\})\right)\right)^{-1/2}. \tag{2.2.39}$$

Proof

Take $L = l$ in (2.2.33), for instance, and then $l \downarrow 0$. ◇

The following result is also a consequence of Theorem 2.2.3.

PROPOSITION 2.2.15

Suppose that the variance σ_n^2 of μ_n, $n \in N^*$, exists. Put $\sigma_{(n)}^2 = \sum_{k=1}^{n} \sigma_k^2$, $n \in N^*$. Then

$$\sum_{k=1}^{n} (1 - Q_{\mu_k}(2l\sigma_{(n)})) \leq A_5 < \infty \tag{2.2.40}$$

for every $n \in N^*$, $l \geq 1 + \alpha$, with $\alpha > 0$ and A_5 is a positive absolute constant depending only on α.

Proof

We have by (2.2.33) and by (1.6.1)

$$1 - 1/l^2 \leq Q_{\mu_{(n)}}(2l\sigma_{(n)}) \leq A\left(\sum_{k=1}^{n} (1 - Q_{\mu_k}(2l\sigma_{(n)}))\right)^{-1/2}$$

so that

$$\sum_{k=1}^{n} (1 - Q_k(2l\sigma_{(n)})) \leq A^2(1 - 1/l^2)^{-2} \leq A_5 < \infty$$

for $l \geq 1 + \alpha$ with $\alpha > 0$. ◇

2.2.4 Most of the theorems obtained in the previous sections have more or less straightforward multidimensional generalizations. We will confine ourselves to spherical multidimensional versions of Theorems 2.2.4 and 2.2.6 in r-dimensional Euclidean space and to some of their consequences.

Let us first state a generalization of the right-hand side inequality (2.2.1) of Lemma 2.2.1.

LEMMA 2.2.16

There is a positive absolute constant $A(r)$, depending only on r, such that

$$Q^{\bigcirc}_{\mu_{(n)}}(l) \leq A(r)a^{-r}\int_{|t| \leq a} |\psi_\mu(t)|\, dt, \tag{2.2.41}$$

where a is a parameter satisfying $0 < al \leq 1$ but otherwise arbitrary.

Proof

We introduce the auxiliary functions

$$U(x) = 2^r\pi^{r/2}\Gamma(1 + r/2)|x|^{-r}(J_{r/2}(|x|/2))^2,$$

where $J_{r/2}$ is the Bessel function of order $r/2$†, and

$$u(t) = (2\pi)^{-r}\int_{R^r} e^{-i(t,x)}U(x)\, dx.$$

† For details on Bessel functions, see, e.g., Tranter (1968).

Then u is a function of $|\boldsymbol{t}|$, $u(\boldsymbol{0}) = 1$, $0 \leq u(\boldsymbol{t}) \leq 1$ for all $\boldsymbol{t} \in R^r$ and $u(\boldsymbol{t}) = 0$ for $|\boldsymbol{t}| \geq 1$ [for a proof of these properties of U and u, see for instance Esséen (1945, p. 101)]. From

$$U(\boldsymbol{x}) = \int_{R^r} \cos(\boldsymbol{t}, \boldsymbol{x}) u(\boldsymbol{t})\, d\boldsymbol{t} \geq \int_{R^r} (1 - \tfrac{1}{2}|\boldsymbol{x}|^2 |\boldsymbol{t}|^2) u(\boldsymbol{t})\, d\boldsymbol{t}$$
$$\geq U(\boldsymbol{0})(1 - \tfrac{1}{2}|\boldsymbol{x}|^2),$$

we see that

$$U(\boldsymbol{x}) \geq \tfrac{1}{2} U(\boldsymbol{0}) = 2^{-r-1} \pi^{r/2} / \Gamma(1 + r/2) \tag{2.2.42}$$

for $|\boldsymbol{x}| \leq 1$.

Let $B \in \mathscr{B}^r$. Consider the easily proved relation

$$\int_{R^r} U(a(\boldsymbol{x} - \boldsymbol{z}))\, d\mu(\boldsymbol{x}) = a^{-r} \int_{|\boldsymbol{t}| \leq a} \psi_\mu(\boldsymbol{t}) u(\boldsymbol{t}/a) e^{-i(\boldsymbol{t}, \boldsymbol{z})}\, d\boldsymbol{t}, \tag{2.2.43}$$

where a is an arbitrary positive parameter and $\boldsymbol{z} \in R^r$. Using (2.2.43) and (2.2.42) and proceeding as in the one-dimensional case we find that

$$Q_\mu^{\bigcirc}(l) \leq 2(U(\boldsymbol{0}))^{-1} a^{-r} \int_{|\boldsymbol{t}| \leq a} |\psi_\mu(\boldsymbol{t})|\, d\boldsymbol{t}$$

for $0 < al \leq 1$ and the lemma is proved. ◇

Let μ_k be a p.m. on $\mathscr{B}^r$, $1 \leq k \leq n$, and let the quantity χ_k be defined by

$$\chi_k(u) = \inf_{|\boldsymbol{t}| = 1} \int_{|\boldsymbol{x}| \leq u} (\boldsymbol{t}, \boldsymbol{x})^2\, d\mu_k^s(\boldsymbol{x}), \tag{2.2.44}$$

i.e. χ_k is the least eigenvalue (possibly zero) of the nonnegative quadratic form $\int_{|\boldsymbol{x}| < u} (\boldsymbol{t}, \boldsymbol{x})^2\, d\mu_k^s(\boldsymbol{x})$ of the variables $t_1, \ldots t_r$. Obviously,

$$\int_{|\boldsymbol{x}| \leq u} (\boldsymbol{t}, \boldsymbol{x})^2\, d\mu_k^s(\boldsymbol{x}) \geq \chi_k(u) |\boldsymbol{t}|^2 \tag{2.2.45}$$

and χ_k is a nondecreasing function for $u > 0$, $1 \leq k \leq n$. If μ_k is nonsingular, it is easily seen that $\chi_k(u) > 0$ for sufficiently large u, $1 \leq k \leq n$.

With the above notations we state the following generalization of Theorem 2.2.4.

THEOREM 2.2.17

For any positive $l_1, \dots, l_n \le L$ we have

$$Q^{\bigcirc}_{\mu(n)}(L) \le A(r)L^r \left(\sum_{k=1}^{n} l_k^{2r}(\mu_k^s(S^c(0; l_k)) + l_k^{-2}\chi_k(l_k)) \right)^{-1/2}, \tag{2.2.46}$$

where $A(r)$ is a positive absolute constant depending only on r.

Proof

Since the method of proof is similar to that used in proving Lemma 2.2.2 and Theorem 2.2.4 we confine ourselves to the main parts of the proof. From Lemma 2.2.16 we get

$$Q^{\bigcirc}_{\mu(n)}(L) \le A(r)L^r \int_{|t| \le 1/L} \exp\left\{ -\tfrac{1}{2} \sum_{k=1}^{n} \int_{R^r} (1 - \cos(t, x))\, d\mu_k^s(x) \right\} dt.$$

For $|t| \le 1/L$

$$\int_{R^r} (1 - \cos(t, x))\, d\mu_k^s(x) \ge \tfrac{11}{24}\chi_k(l_k)|t|^2 + \int_{|x| \ge l_k} (1 - \cos(t, x))\, d\mu_k^s(x),$$

where $\chi_k(l_k)$ is defined by (2.2.44), $1 \le k \le n$. Introducing the quantities $p_k = \int_{|x| \ge l_k} d\mu_k^s(x)$ [where without any loss of generality $\chi_k(l_k) > 0$ and $p_k > 0$],

$$\gamma = \sum_{k=1}^{n} (l_k^{2r} p_k + l_k^{2r-2}\chi_k(l_k)), \tag{2.2.47}$$

$$\beta_k = l_k^{2r-2}\chi_k(l_k)/\gamma, \qquad \gamma_k = l_k^{2r} p_k/\gamma, \qquad 1 \le k \le \text{n},$$

and the p.m.'s

$$dN_k(x) = \begin{cases} 0 & \text{if } |x| < l_k, \\ d\mu_k^s(x)/p_k & \text{if } |x| \ge l_k, \end{cases}$$

we get from Hölder's inequality

$$(2.2.48)\quad Q^{\bigcirc}_{\mu(n)}(L) \le A(r)L^r \prod_{k=1}^{n}\left(\int_{|t|\le 1/L} \exp\{-\tfrac{1}{4}\tfrac{1}{8}\, l_k^{2-2r}\gamma|t|^2\}\,dt\right)^{\beta_k}$$
$$\times \prod_{k=1}^{n}\left(\int_{|t|\le 1/L} \exp\left\{-\tfrac{1}{2}l_k^{-2r}\gamma\int_{R^r}(1-\cos(t,x))\, dN_k(x)\right\}dt\right)^{\gamma_k}$$
$$= A(r)L^r \prod_{k=1}^{n} I_k^{\beta_k}\prod_{b=1}^{n} T_k^{\gamma_k}$$

Now

$$(2.2.49)\quad I_k \le (2/L)^{r-1}\int_{|t|\le 1/L} \exp\{-\tfrac{1}{4}\tfrac{1}{8}\, l_k^{2-2r}\,\gamma t_1^2\}\,dt_1$$
$$\le A(r)(l_k/L)^{r-1}\gamma^{-1/2} \le A(r)\gamma^{-1/2}, \qquad 1\le k\le n.$$

We apply Jensen's inequality to T_k and obtain

$$T_k \le \int_{|x|\ge l_k}\left(\int_{|t|\le 1/L}\{-\tfrac{1}{2}l_k^{-2r}\gamma(1-\cos(t,x))\}dt\right) dN_k(x), \qquad 1\le k\le n.$$

Denote the inner integral by I_k'. Since I_k' depends on only $|x|$ (and L), we may choose $x = (|x|, 0, \ldots, 0)$, and we get

$$I_k' = \int_{|t|\le 1/L} \exp\{-\tfrac{1}{2}l_k^{-2r}\gamma(1-\cos(|x|t_1))\}\,dt_1\cdots dt_r$$
$$\le (2/L)^{r-1}|x|^{-1}\int_{|u|\le |x|/L} \exp\{-\tfrac{1}{2}l_k^{-2r}\gamma(1-\cos u)\}\,du.$$

Since $|x|\ge l_k$ and $l_k \le L$ we get by the argument in (2.2.17)

$$I_k' \le A(r)\gamma^{-1/2}, \qquad 1\le k\le n,$$

and hence

$$(2.2.50)\qquad T_k \le A(r)\gamma^{-1/2}, \qquad 1\le k\le n.$$

From (2.2.48)–(2.2.50) and the definition (2.2.47) of γ the proposed inequality follows.† ◇

From Theorem 2.2.17 we get immediately the following consequences.

† In the above proof the $A(r)$'s may be different. However, they are positive absolute constants depending only on r.

PROPOSITION 2.2.18

For any positive $l_1, \ldots, l_n \leq L$ we have

$$Q^{\bigcirc}_{\mu(n)}(L) \leq A(r)L^r\left(\sum_{k=1}^{n} l_k^{2r}(1 - Q^{\bigcirc}_{\mu_k}(l_k)\right)^{-1/2} \tag{2.2.51}$$

where $A(r)$ is a positive absolute constant depending only on r.

Proposition 2.2.18 is the spherical multidimensional version of Proposition 2.2.9.

PROPOSITION 2.2.19

For any positive $l \leq L$ we have

$$Q^{\bigcirc}_{\mu(n)}(L) \leq A(r)(L/l)^r\left(\sum_{k=1}^{n} (1 - Q^{\bigcirc}_{\mu_k}(l))\right)^{-1/2} \tag{2.2.52}$$

where $A(r)$ is a positive absolute constant depending on r. If, furthermore, $l = L$ and $\mu_1 = \cdots = \mu_n = \mu$ with μ nondegenerate, then

$$Q^{\bigcirc}_{\mu(n)}(l) \leq A(r; \mu; l)n^{-1/2}, \tag{2.2.53}$$

where $A(r; \mu; l)$ is a positive constant independent of n but depending on r, μ, and on l.

Proposition 2.2.19 is the spherical multidimensional version of Propositions 2.2.10 and 2.2.11.

Of course, we might continue to get spherical multidimensional versions. Instead, let us give an application of (2.2.52) in order to get a single proof of the generalization of Theorem 2.1.5.

PROPOSITION 2.2.20

We have $Q^{\bigcirc}_{\mu * \nu} = Q^{\bigcirc}_{\mu}$ iff $\nu = \delta_a$.

Proof

In one sense the proof is trivial. Hence assume that $Q^{\bigcirc}_{\mu * \nu} = Q^{\bigcirc}_{\mu}$.

Then (cf. p. 40), $Q_\mu^\bigcirc \le Q_\nu^\bigcirc$ and $Q_\mu^\bigcirc \le Q_{\nu(n)}^\bigcirc$ for every $n \in N^*$. If $\nu \ne \delta_a$, then there is an $l > 0$ such that $0 < Q_\nu^\bigcirc(l) < 1$. Using (2.2.52) ,we obtain

$$(2.2.54) \qquad Q_{\nu(n)}^\bigcirc(l) \le A(r)(n(1 - Q_\nu^\bigcirc(l)))^{-1/2},$$

so that, for sufficiently large n, we get from (2.2.54) $Q_{\nu(n)}^\bigcirc(l) < Q_\mu^\bigcirc(l)$. Hence we are led to a contradiction, and consequently $\nu = \delta_a$. ◇

The above result also holds for $Q_\mu^\square$. In fact, it suffices to apply the rectangular analogue of (2.2.52). However, simple examples show that this property is not true for $Q_\mu^\urcorner$.

Let us suppose now for a moment that $\mu_1 = \cdots = \mu_n = \mu \ne \delta_\alpha$. It follows basically from (2.2.53) that $Q_{\mu(n)}^\bigcirc(l) = O(n^{-1/2})$ as $n \to \infty$. This order of magnitude cannot be improved, at least not in the general case, indeed, take $r = 2$ and suppose that μ is a p.m. equally distributed at the mass points (1, 1) and $(-1, -1)$. Then

$$\mathop{\ast}_{k=1}^{2n} \mu_k(\{0\}) = \binom{2n}{n} 2^{-2n} \sim (\pi n)^{-1/2}.$$

If, however the μ_k are not singular, $Q_{\mu(n)}^\bigcirc(l)$ should be of order $n^{-r/2}$. From an inequality which well be stated in the next theorem it follows the $Q_{\mu(n)}^\bigcirc(l)$ is in fact of order $n^{-r/2}$ as $n \to \infty$. This inequality is the spherical multidimensional version of (2.2.28) of Theorem 2.2.6. From now on we no longer assume that $\mu_k = \mu$, $1 \le k \le n$.

THEOREM 2.2.21

For any positive $l \le L$ we have

$$(2.2.55) \qquad Q_{\mu(n)}^\bigcirc(L) \le A(r)(L/l)^r \left(\sup_{u \ge l} u^{-2} \sum_{k=1}^{n} \chi_k(u) \right)^{-r/2},$$

where $A(r)$ is a positive absolute constant depending only on r.

Proof

Since for any $L > l$ we have $Q_\mu^\bigcirc(L) \le \text{const}\,(L/l)^r Q_\mu^\bigcirc(l)$, it is sufficient to prove (2.2.55) for $l = L$. From Lemma 2.2.16 we get

$$Q^{\bigcirc}_{\mu_{(n)}}(l) \le A(r)a^{-r}\int_{|t|\le a}\exp\left\{-\tfrac{1}{2}\sum_{k=1}^{n}\int_{R^r}(1-\cos(t,x))\,d\mu_k^s(x)\right\}dt,$$

where $0 < al \le 1$. Using

$$\int_{R^r}(1-\cos(t,x))\,d\mu_k^s(x) \ge \int_{|x|\le 1/|t|}(1-\cos(t,x))\,d\mu_k^s(x)$$
$$\ge \tfrac{1}{2}\tfrac{1}{4}\int_{|x|\le 1/|t|}(t,x)^2\,d\mu_k^s(x) \ge \tfrac{1}{2}\tfrac{1}{4}\chi_k(1/|t|)|t|^2, \qquad 1 \le k \le n,$$

we have

$$Q^{\bigcirc}_{\mu_{(n)}}(l) \le A(r)a^{-r}\int_{|t|\le a}\exp\left\{-\tfrac{1}{4}\tfrac{1}{8}\sum_{k=1}^{n}\chi_k(1/|t|)|t|^2\right\}dt.$$

Since χ_k is a nondecreasing function of $u > 0$, $1 \le k \le n$, it follows that

$$Q^{\bigcirc}_{\mu_{(n)}}(l) \le A(r)a^{-r}\left(\sum_{k=1}^{n}\chi_k(1/a)\right)^{-r/2}$$

Putting $a = u^{-1}$ and observing that $u \ge l$ we get the desired inequality. ◇

Clearly, we get

COROLLARY

For any positive $l \le L$ and for $\mu_1 = \cdots = \mu_n = \mu$ with μ nonsingular, we have

$$Q^{\bigcirc}_{\mu_{(n)}}(L) \le A(r)(L/l)^r\left(\sup_{u\ge l} u^{-2}\chi_1(u)\right)^{-r/2} n^{-r/2},$$

where $A(r)$ is a positive absolute constant depending only on r.

Complements

1 For any positive $l_1, \ldots, l_n \ge L$ we have

$$Q_{\mu_{(n)}}(L) \le AL\max_{1\le k\le n} Q_{\mu_k}(L)\left(\sum_{k=1}^{n} l_k^2 D^2(\mu_k^s; l_k)\right)^{-1/2}$$

where A is a positive absolute constant [Kesten (1969); see also Kesten (1972)].

2 Suppose that the conditions of Proposition 2.2.13 are fulfilled and set $h_{n_0} = \sup\{h: Q_{\mu(n)}(hln^{1/2}) \leq \eta\}$. Then $h' = \lim_{n_0 \to \infty} h_{n_0}$ exists and equals $\sigma_{\mu_0} Z_\lambda(\eta)/l$, where $\sigma^2_{\mu_0}$ is defined by (1.6.2) with a prelaced by $1 - \varepsilon$, and Z_λ is the ds.f. of the normal p.m. with mean 0 and variance 1 [Lévy (1937; 1954, p. 156–157); see also Doeblin and Lévy (1936)].

3 Suppose that the p.m. μ_k has expectation zero, variance σ_k^2 and absolute moment of third order β_k^3, $1 \leq k \leq n$, and let $\kappa = \min_{1 \leq k \leq n} \sigma_k/\beta_k$. Then

$$Q_{\mu(n)}(l) \leq \frac{3072 \ln n}{\kappa^9 n^{1/2}} \cdot \left\{\ln n + \frac{\kappa^3 l}{8 \min_{1 \leq k \leq n} \sigma_k}\right\}$$

[Offord (1945, Theorem 1, p. 468); see also Dunnage (1971, Theorem A, p. 489)].

4 In Esséen (1966, Theorem 2, p. 214), there is an inequality similar to (2.2.51) stated for rectangular cn.f.'s and with the right-hand side depending on the cn.f.'s defined with respect to certain unbounded domains. In this respect (2.2.51) is more satisfactory than the older one because it concerns only spherical cn.f.'s. From (2.2.51) a Kolmogorov type inequality for rectangular cn.f.'s can easily be obtained.

5 We have seen that a spherical multidimensional version of Proposition 2.2.11 may be obtained from Theorem 2.2.17. Another possibility which involves the cn.f.'s of order u associated with a p.m. (see Definition 1.7.7) was indicated by Sazonov (1966), who examined Sebast'yanov's (1963) conjecture: if $\mu_k = \mu$, $1 \geq k \leq n$, with μ a nondegenerate p.m. on $\mathscr{B}^r$, then for any $1 \leq u \leq r$

$$Q_{\mu(n),u}(l) \leq A(\mu; l) n^{-(u-r+1)/2}$$

where $A(\mu; l)$ is a positive constant independent on n and u but depending on μ and on l. He proved that Sebast'yanov's conjecture is true when some additional restrictions are assumed on μ [see also Prohorov and Rozanov (1969, p. 178)]. However, an explicit expression of the dependence of $A(\mu; l)$ on μ and on l is not given. In Theorems 2.2.17 and 2.2.21 such an explicit expression has been obtained, even in the case when the μ_k are not all equal, but on the other hand only spherical cn.f.'s were considered.

Notes and Comments

The uniform distance between the d.f. $F_{(n)}$ associated with $\mu_{(n)}$ and infinitely divisible d.f.'s was studied by Kolmogorov (1956, 1963), and between $F_{(n)}$ and suitable chosen Poisson exponentials by Lecam (1965a, b). In these investigations a certain inequality for $Q_{\mu_{(n)}}$ plays an important part. The original Kolmogorov version of this inequality, stated in (1956) and proved in (1958), was later improved and generalized by Rogozin (196lb). He combined Kolmogorov's methods with combinatorial arguments to prove (2.2.32). A somewhat more general Kolmogorov-type inequality is (2.2.29), from which (2.2.30) easily follows. A proof of (2.2.30) in the case $l_1 = \cdots = l_n = l$ can be found in Lecam (1965a), where Kolmogorov's method of proof is used. Esséen (1966), using c.f.'s, was able to give a new proof of (2.2.32). Later, Esséen (1968) showed that the inequalities (2.2.30) and (2.2.32) are consequences of (2.2.29), which in turn follows from the main inequality (2.2.23). In the present text Esséen's (1968) proof of (2.2.23) is given; it is essentially based on Lemma 2.2.1, the proof of which uses a certain convolution method which was applied in Esséen (1966) and earlier by Rosén (1961) to obtain the upper bound. Moreover a rather general inequality [(2.2.10), Lemma 2.2.2] is proved by means of which the upper bound in Lemma 2.2.1 can be estimated; this result is stated in such a way that it can be immediately applied to an infinitely divisible p.m. These two lemmas lead first to a new proof of the inequlity (2.2.19) (Theorem 2.2.3) for the cn.f. of an infinitely divisible p.m. obtained previously by Lecam (1965a), as well as to a new proof of a result by Doeblin (1939)

(Corollary of Theorem 2.2.3). Next, these lemmas lead to the main inequality (2.2.23). A sharper form of the Kolmogorov–Rogozin (see Complement 1, p. 69) inequality (2.2.32) was given by Kesten (1969) [see also Kesten (1972)]; the proof is essentially based on (2.2.32) and combinational methods.

Special forms of the main inequality (2.2.33) were previously given by Rogozin (1961a) [see (2.2.33)], Rosén (1961) [see (2.2.35); however, no explicit determination of the constant $A(\mu; l)$ and its dependence on μ and l is given in Rosén's paper], Doeblin (1939) [see (2.2.37)], Lévy (1937, 1954) [see (2.2.38)] [see also Doeblin and Lévy (1936)], Rogozin (1961a) [see (2.2.39)]. Doeblin and Lévy (1936) were the first to give quantitative estimates for the spreading out of $\mu_{(n)}$ by means of cn.f.'s.

Inequalities more primitive than those quoted above were obtained assuming existence of moments up to the third order. Such inequalities were among the first to be found; they go back to Littlewood and Offord (1943) who gave an upper bound for $Q_{\mu_{(n)}}$ when studying the number of real roots of a random algebraic equation. Their inequality was sharpened by Erdös (1945) and extended in more general case by Offord (1945) (see Complement 3, p. 70) A thorough study of such inequalities is to be found in Dunnage (1971) [see also Dunnage (1968)].

The multidimensional versions described in Section 2.2.4 are all due to Esséen (1968); the application mentioned in Proposition 2.2.20 is due to Hengartner and Theodorescu (1972a). Multidimensional versions assuming existence of moments up to the third order are to be found in Dunnage (1971).

2.3 IDENTICAL FACTORS

2.3.1 Let us examine now more in detail the case $\mu_1 = \cdots = \mu_n = \mu$ with μ nondegenerate. We have already obtained several results, e.g., the corollary of Theorem 2.2.6 and the second part of Proposition 2.2.11.

Set [cf. (2.2.27)]

$$\chi(t) = \int_{|x| \leq 1/|t|} x^2 \, d\mu^s(x); \tag{2.3.1}$$

evidently χ is an even nonincreasing function for $t > 0$, and it is easily seen that $\lim_{t \to 0} \chi(t) < \infty$ iff $\int_R x^2 \, d\mu(x) < \infty$.

THEOREM 2.3.1

Suppose that $\int_R x^2 d\mu(x) = \infty$. Then for every fixed $l > 0$ we have $Q_{\mu_{(n)}}(l) = o(n^{-1/2})$ as $n \to \infty$.

Proof

From Lemma 2.2.5 we find that

$$Q_{\mu_{(n)}}(l) \leq Aa^{-1} \int_{|t| \leq a} \exp\{-\tfrac{1}{4}\tfrac{1}{8} nt^2 \chi(t)\} \, dt, \tag{2.3.2}$$

where a is fixed but so small that $0 < al \leq 1$ and $\chi(a) > 0$. This is always possible since μ is nondegenerate.

Let us now suppose that $\int_R x^2 \, d\mu(x) = \infty$, i.e. $\lim_{t \to 0} \chi(t) = \infty$. Then for $0 < \varepsilon < a$ we get from (2.3.2)

$$\begin{aligned} Q_{\mu_{(n)}}(l) &\leq Aa^{-1} \int_{|x| \leq \varepsilon} \exp\{-\tfrac{1}{4}\tfrac{1}{8} nt^2 \chi(\varepsilon)\} \, dt \\ &\quad + Aa^{-1} \int_{\varepsilon}^{a} \exp\{-\tfrac{1}{4}\tfrac{1}{8} nt^2 \chi(a)\} \, dt \\ &\leq \theta_1 (n\chi(\varepsilon))^{-1/2} + \theta_2 n^{-1/2} \int_{\varepsilon(n\chi(a))^{1/2}} \exp\{-\tfrac{1}{4}\tfrac{1}{8} u^2\} \, du, \end{aligned}$$

where θ_1 and θ_2 are constants not depending on n or on ε. Choosing $\varepsilon = n^{-1/4}$ we have $\lim_{n \to \infty} \chi(n^{-1/4}) = \infty$ and thus $Q_{\mu_{(n)}}(l) = o(n^{-1/2})$ as $n \to \infty$. ◇

It is seen from the proof of Theorem 2.3.1 that the faster the integral $\int_R x^2\, d\mu^s(x)$ diverges the faster will $Q_{\mu(n)}(l)$ tend to zero as $n \to \infty$. This observation is confirmed by the next theorem.

THEOREM 2.3.2

Suppose that $\beta_r = \int_R |x|^r\, d\mu(x) < \infty$, where r is a constant and $0 < r \leq 2$. Then

$$Q_{\mu(n)}(l) \geq K(r)l(l + (n\beta_r(a))^{1/r})^{-1}, \tag{2.3.3}$$

where $\beta_r(a) = \int_R |x - a|^r\, d\mu(x)$ and a arbitrary. Moreover, the constant $K(r)$ depends only on r and may be given the value

$$K(r) = \begin{cases} \frac{1}{4}r(r+1)^{-(1+1/r)} & \text{if } 0 < r < 2, \\ (54)^{-1/2} & \text{if } r = 2. \end{cases}$$

Proof

We restrict ourselves to the case $0 < r < 2$.

Let us write $\beta_r^s = \int_R |x|^r\, d\mu^s(x)$ and $\mu_{(n)}^s = \mathop{\ast}\limits_{k=1}^{n} \mu_k^s$. Applying an inequality proved by von Bahr and Esséen (1965), Theorem 1, to symmetrical p.m.'s, we get

$$\int_R |x|^r\, d\mu_{(n)}^s(x) \leq n\beta_r^s$$

and hence from the Markov inequality

$$\mu_{(n)}^s([-(kn\beta_r^s)^{1/r}, (kn\beta_r^s)^{1/r}]) \geq 1 - k^{-1},$$

where $k > 1$. Thus

$$Q_{\mu^s(n)}(2(kn\beta_r^s)^{1/r}) \geq 1 - k^{-1}.$$

Since

$$Q_{\mu^s(n)}(2(kn\beta_r^s)^1 1,) \leq (2l^{-1}(kn\beta_r^s)^{1/r} + 1)Q_{\mu(n)}^s(l),$$

we get

$$Q_{\mu_{(n)}}(l) \geq Q_{\mu^s_{(n)}}(l) \geq l(l + 2(kn\beta_r^s)^{1/r})^{-1}(1 - k^{-1}).$$

But

$$\beta_r^s \leq 2^r \int_R |x - a|^r \, d\mu(x),$$

whence

$$Q_{\mu_{(n)}}(l) \geq \tfrac{1}{4}\, l(l + (n\beta_r(a))^{1/r})^{-1}k^{-1/r}(1 - k^{-1}).$$

For $k = r + 1$, the function $k^{-1/r}(1 - k^{-1})$ is as large as possible, whence the constant $K(r)$ of the theorem.

If $r = 2$, we proceed similarly but apply Chebychef's inequality. ◇

Let us remark that for moderately large l it may be preferable to write (2.3.3) in the form

$$Q_{\mu_{(n)}}(l) \geq K(r)l(l + (\beta_r(a))^{1/r})^{-1}n^{-1/r}, \tag{2.3.4}$$

valid for all l but less suitable if l is large. Moreover, if μ has a finite variance σ_μ^2, then we have

$$Q_{\mu_{(n)}}(l) \geq (54)^{-1/2}l(l + \sigma_\mu n^{1/2})^{-1} \geq (54)^{-1/2}l(l + \sigma_\mu)^{-1}n^{-1/2}. \tag{2.3.5}$$

2.3.2 The next result is an immediate consequence of Theorems 2.3.1, 2.3.2, and 2.2.4.

THEOREM 2.3.3

There exist positive constants $K_1(\mu; l)$ and $K_2(\mu; l)$ only depending on μ and on l such that

$$K_1(\mu, l)n^{-1/2} \leq Q_{\mu_{(n)}}(l) \leq K_2(\mu; l)n^{-1/2}$$

iff $\int_R x^2 \, d\mu(x) < \infty$.

Note that the right-hand part of the above inequality is in fact given by (2.2.35).

Complement

If μ is a nondegenerate p.m. belonging to the normal domain of attraction of a stable p.m., then inequalities similar to those given in Theorem 2.3.3, and containing the exponent of μ, hold true [Esséen (1968, pp. 300–303)].

Notes and Comments

All the results of this section are due to Esséen (1968).

2.4 ASYMPTOTIC ESTIMATIONS

2.4.1 We discuss here several other consequences of Lemma 2.2.1 which lead to asymptotic estimations for cn.f.'s of convolutions.

Let $(\mu_n)_{n \in N^*}$ be a sequence of p.m.'s. We shall say that this sequence satisfies *Condition* Δ if there exist a constant $\eta > 0$ and a real-valued function κ defined on N^* such that $\kappa(n) \to \infty$ and

$$\int_{|t| \leq \eta} \prod_{k=1}^{n} |\psi_{\mu_k}(t)| \, dt \leq B_0/\kappa(n) \tag{2.4.1}$$

for every $n \geq n_0$, where n_0 and B_0 are positive constants.

Condition Δ is satisfied for $\kappa(n) = n^{1/2}$ and for $\mu_n = \mu \neq \delta_\alpha$, $n \in N^*$. Indeed by Rosén (1961, p. 323), there exist two positive constants η and c such that

$$|\psi_\mu(t)| \leq 1 - ct^2$$

for every $|t| \leq \eta$; it follows then that there exist two positive constants η and γ satisfying

$$|\psi_\mu(t)| \leq e^{-\gamma t^2} \tag{2.4.2}$$

for every $|t| \leq \eta$. Therefore,

$$\int_{|t| \leq \eta} |\psi_\mu(t)|^n \, dt \leq \int_{|t| \leq \eta} \exp(-n\gamma t^2) \, dt$$
$$\leq n^{-1/2} \int_R \exp(-\gamma u^2) du = Bn^{-1/2}$$

for all $n \in N^*$.

We shall find a more general condition which is sufficient in order that Condition Δ be satisfied.

LEMMA 2.4.1

Let $(\mu_n)_{n \in N^*}$ be a sequence of p.m.'s and suppose that there are a nondegenerate p.m. μ and an $\varepsilon > 0$ such that $|\psi_{\mu_k}(t)| \leq |\psi_\mu(t)|$ holds for every $|t| \leq \varepsilon$ and for infinitely many k. Denote by $\nu(n)$ the number of μ_k, $k \leq n$, satisfying this condition. Then the sequence $(\mu_n)_{n \in N^*}$ satisfies Condition Δ with $x(n) = (\nu(n))^{1/2}$.

Proof

In fact there exist two positive constants η and γ such that (2.4.2) holds. Putting $\eta_0 = \min(\varepsilon, \eta)$, we have

$$\int_{|t| \leq \eta_0} \prod_{k=1}^{n} |\psi_{\mu_k}(t)| \, dt \leq \int_{|t| \leq \eta_0} |\psi_\mu(t)|^{\nu(n)} \, dt$$
$$\leq \int_{|t| \leq \eta_0} \exp(-\nu(n)\gamma t^2 \, dt$$
$$\leq (\nu(n))^{1/2} \int_R \exp(-\gamma u^2) \, du = \mathrm{B}_0 \, (\nu(n))^{-1/2},$$

i.e. (2.4.1). Thus Condition Δ is satisfied or $\kappa(n) = (\nu(n))^{1/2}$. ◇

If $\liminf_{n \to \infty} \nu(n)/n > 0$, then Condition Δ is satisfied for $\kappa(n) = n^{1/2}$.

In the following we need an auxiliary result which is obtained by modifying slightly the right-hand side for (2.2.1).

LEMMA 2.4.2

For any nonnegative l there is a positive absolute constant A_2 such that

$$Q_\mu(l) \leq A_2 \max(l, a^{-1}) \int_{|t| \leq a} |\psi_\mu(t)|\, dt, \tag{2.4.3}$$

where a is an arbitrary positive parameter.

Proof

Let us adapt the proof of Lemma 2.2.1. We start with (2.2.7) which holds for $0 < al \leq 1$. If we put in (2.2.7) $a = l^{-1}$, then we get

$$Q_\mu(l) \leq A_2 l \int_{|t| \leq 1/l} |\psi_\mu(t)|\, dt \tag{2.4.4}$$

for every $l > 0$. Consequently, for every positive a and l such that $al \leq 1$, (2.4.3) follows from (2.2.7). In case $al > 1$, (2.4.4) becomes

$$Q_\mu(l) \leq A_2 l \int_{|t| \leq a} |\psi_\mu(t)|\, dt. \tag{2.4.5}$$

The evaluation (2.4.3) with $A_2 = (\frac{96}{95})^2$ follows from (2.2.7) and (2.4.5) for any positive a and l. Since Q_μ is nondecreasing, (2.4.3) and (2.2.7) hold also for $l = 0$ for any $a > 0$. ◇

2.4.2 Let us state our main result:

THEOREM 2.4.3

Suppose that the sequence of p.m.'s $(\mu_n)_{n \in N^*}$ satisfies Condition Δ. Then for any $l \geq 0$ and for sufficiently large n we have

$$Q_{\mu(n)}(l) \leq B_1 (l + 1)/\kappa(n), \tag{2.4.6}$$

where B_1 is a constant which does not depend on l and on n.

Proof

Let us apply Lemma 2.4.2 to $\mu_{(n)}$ and with $a = \eta$; we obtain

$$Q_{\mu_{(n)}}(l) \leq A_2 \max(l, \eta^{-1}) \int_{|t| \leq \eta} \prod_{k=1}^{n} |\psi_{\mu_k}(t)| \, dt$$

for any $l \geq 0$. Thus from Condition Δ we get

$$Q_{\mu_{(n)}}(l) \leq A_2 \max(l, \eta^{-1}) B_0/\kappa(n) \tag{2.4.7}$$

and in turn (2.4.7) yields (2.4.6). ◇

Clearly the estimate in (2.4.7) is more precise than (2.4.6).

From Theorem 2.4.3 we get the following simple

COROLLARY

Suppose that $\mu_1 = \cdots = \mu_n = \mu$ with μ nondegenerate. Then for all $l \geq 0$ we have

$$Q_{\mu_{(n)}}(l) \leq B_2 (l + 1) n^{-1/2}, \tag{2.4.8}$$

where B_2 is a constant which does not depend on l and on n.

Note that we can take $B_2 = (\frac{9}{9}\frac{6}{5})^2 \pi^{1/2}/\min(1, \eta)\gamma^{1/2}$ in (2.4.8), where η and γ are indicated in (2.4.2).

Let us now consider the case when l depends on n. We obtain immediately

THEOREM 2.4.4

Suppose that the sequence of p.m.'s $(\mu_n)_{n \in N^*}$ satisfies Condition Δ. Then as $n \to \infty$ we have

(a) $Q_{\mu_{(n)}}(l_n) = O((l_n + 1)/\kappa(n))$;

(b) $Q_{\mu_{(n)}}(l_n) = o(1)$ if $l_n = o(\kappa(n))$;

(c) $Q_{\mu_{(n)}}(l_n) = O(1/\kappa(n))$ if $l_n = O(1)$;

(d) $Q_{\mu_{(n)}}(l_n) = O(n^{p-1/2})$ if $\kappa(n) = n^{1/2}$ and $l_n = O(n^p)$, $0 < p < \frac{1}{2}$;

(e) $\sup_{x \in R} \mu_{(n)}(\{x\}) = O(1/\kappa(n))$.

Using examples mentioned by Rosén (1961, p. 327), and by Esséen (1945, p. 53), it is easy to show that it is not possible to improve the above evaluations. As an example, take (d), Theorem 2.4.4; if $\mu_k = \mu$, $k \in N^*$, is the normal p.m. with mean 0 and variance 1, then

$$Q_{\mu(n)}(2n^p)\, n^{p-1/2} \to (2/\pi)^{1/2}.$$

This shows that (d) cannot be generally improved.

Notes and Comments

The results of this section are due to Petrov (1970). The proof of Lemma 2.4.2 is a straightforward modification of the proof of Lemma 2.2.1 due to Esséen (1966, 1968). Theorem 2.4.4 is a generalization of some results of Chung and Erdös (1951), Rosén (1961), and Heyde (1966).

3 CONVERGENCE PROBLEMS

The aim of this chapter is to show the use of cn.f.'s in proving convergence theorems. The idea is basically due to Lévy who applied it systematically in his monograph (1937, 1954). It is closely connected with centering at medians and symmetrization, and consequently is relatively recent, while centering at expectations goes back to Bernoulli. Centering at medians has certain advantages over other centering procedures, e.g., medians always exist. Section 3.1 has an auxiliary character, providing results to be further used in proving convergence theorems. Section 3.2 deals with convergence theorems. Section 3.3 examines the main results concerning essential convergence.

3.1 AUXILIARY RESULTS

3.1.1 Consider the space $\mathfrak{F}$ and let us define a mapping d^{L} from $\mathfrak{F} \times \mathfrak{F}$ to R as follows:

(3.1.1)
$$d^{\mathrm{L}}(F, G) = \inf\{\varepsilon : F(x - \varepsilon) - \varepsilon \leq G(x) \leq F(x + \varepsilon) + \varepsilon \quad \text{for every } x \in R\}.$$

The function d^{L} so defined is a distance, and $(\mathfrak{F}, d^{\mathrm{L}})$ is a complete metric space. Geometrically, $2^{1/2}d^{\mathrm{L}}(F, G)$ is the maximal Euclidean distance between F and G in the direction, of the second bisector, i.e., in the direction $(-1, 1)$. More precisely, let Γ_F be the continuous curve containing the set $(x, F(x))$ and the vertical segments joining $(x, F^-(x))$ and $(x, F(x))$; then corresponding to each point $(x, y) \in \Gamma_F$ there exists a unique point $(x - t, y + t) = (a(x, y), b(x, y)) \in \Gamma_G$ such that

$$d^{\mathrm{L}}(F, G) = 2^{-1/2} \sup_{(x, y) \in \Gamma_F} ((a(x, y) - x)^2 + (b(x, y) - y)^2)^{1/2}. \tag{3.1.2}$$

The following notions are to be used below. We say that a sequence $(F_n)_{n \in N^*}$ of d.f.'s† converges *weakly* to a d.f. F and write $F_n \xrightarrow{\text{w}} F$, if F_n converges to F at every continuity point of F. This definition is justified, i.e., the limit, if it exists, is unique. We say also that a sequence $(F_n)_{n \in N^*}$ of d.f.'s converges *completely* and write $F_n \xrightarrow{\text{c}} F$, if $F_n \xrightarrow{\text{w}} F$ and $F_n(\pm\infty) \to F(\pm\infty)$. Clearly, weak convergence does not imply complete convergence. However, if we restrict ourselves to the Lévy space $\mathfrak{F}$, then these convergences are equivalent.

Next we say that a sequence $(\mu_n)_{n \in N^*}$ of p.m.'s on $\mathscr{B}$ converges weakly to a p.m. μ on $\mathscr{B}$, and write $\mu_n \xrightarrow{\text{w}} \mu$, if $\int_R g(x)\, d\mu_n(x) \to \int_R g(x)\, d\mu(x)$ as $n \to \infty$ for every bounded, continuous, real-valued function g on R.

We quote here the basic Helly theorem [see, e.g., Loève (1963, p. 179)], which asserts that every sequence of d.f.'s is weakly compact.‡ We mention also [see, e.g., Gnedenko and Kolmogorov (1954, p. 33)] that the following four assertions are equivalent for $F_n, F \in \mathfrak{F}$, $n \in N^*$:

(a) $F_n \xrightarrow{\text{w}} F$ as $n \to \infty$;

(b) F_n converges to F as $n \to \infty$ on a set dense in R;

(c) $d^{\mathrm{L}}(F_n, F) \to 0$ as $n \to \infty$;

(d) $\mu_n \xrightarrow{\text{w}} \mu$ as $n \to \infty$, where μ_n and μ are the p.m.'s induced by F_n, $n \in N^*$, respectively by F.

† Sometimes we shall understand by a d.f. F a real-valued function defined on R which is finite, nondecreasing, and right continuous. Obviously, if $F(-\infty) = 0$, $F(+\infty) = 1$, then F is the d.f. of a r.v., i.e., $F \in \mathfrak{F}$.

‡ A set is *compact* with respect to a type of convergence if every infinite sequence in the set contains a subsequence which converges in the same sense.

Finally, we remark that a set $S \subset \mathfrak{F}$ is compact iff $F(x) \to 0$ as $x \to -\infty$ and $F(x) \to 1$ as $x \to \infty$, uniformly on S.

Concerning r.v.'s, we shall deal with a.s. convergence, convergence in probability, convergence in law and convergence in q.m. We shall use the following notations for these convergences, respectively: $\xrightarrow{\text{a.s}}$, $\xrightarrow{\text{p}}$, $\xrightarrow{\text{L}}$, and $\xrightarrow{\text{q.m}}$. Convergence in law for r.v.'s is, by definition, weak convergence for their d.f.'s.

3.1.2 Let us compare now the Lévy distance d^{L} between d.f.'s functions and their corresponding concentration functions.

PROPOSITION 3.1.1

Let $F, G \in \mathfrak{F}$. Then

$$d^{\mathrm{L}}(Q_F, Q_G) \leq 2d^{\mathrm{L}}(F, G). \tag{3.1.3}$$

Proof

Let $d = d^{\mathrm{L}}(F, G)$; then we can write

$$G(x - d) - d \leq F(x) \leq G(x + d) + d$$

and

$$G(x + l - d) - d \leq F(x + l) \leq G(x + l + d) + d$$

so that

$$G(x + l - d) - G(x + d) - 2d \leq F(x + l) - F(x)$$
$$\leq G(x + l + d) - G(x - d) + 2d;$$

we get

$$\sup_{x \in R} (G(x + l - d) - G(x + d)) - 2d \leq Q_F(l)$$
$$\leq \sup_{x \in R} (G(x + l + d) - G(x - d)) + 2d,$$

i.e.,

$$Q_G(l - 2d) - 2d \leq Q_F(l) \leq Q_G(l + 2d) + 2d.$$

Consequently,

$$d^{\mathrm{L}}(Q_F, Q_G) \leq 2d^{\mathrm{L}}(F, G). \quad \diamond$$

Proposition 3.1.1 above shows that the mapping $F \to Q_F$ of $\mathfrak{F}$ into $\mathfrak{Q}$ is continuous and so, by Theorem 1.1.6, we conclude that $F \in \tilde{\mathfrak{F}}_+$ are the fixed points of this mapping.

COROLLARY

Let F, $G \in \mathfrak{F}$, suppose that there exists at least one l_0 such that

$$Q_G(l_0) \leq \alpha < 1. \tag{3.1.4}$$

Then

$$d^{\mathrm{L}}(Q_{F*G}, Q_F) > 0. \tag{3.1.5}$$

Proof

In fact by Theorem 2.1.5, $d^{\mathrm{L}}(Q_{F*G}, Q_F) = 0$ iff $G = H_a$. Consequently, by (3.1.4) it follows that $G \neq H_a$ so that (3.1.5) is proved. ◇

Notes and Comments

The metric space (F, d^{L}) was defined by Lévy (1937, 1954), who used the geometrical interpretation (3.1.2) of d^{L}. Proposition 3.1.1. is due to Sâmboan and Theodorescu (1968), whereas its corollary is to be found in Lévy (1937, 1954). For more general spaces, Prohorov (1956) defined a distance between p.m.'s (see p. 124) which reduces to the Lévy distance for one-dimensional Euclidean space, and which is equivalent to weak convergence of p.m.'s [for details, see Billingsley (1968)].

3.2 CONVERGENCE THEOREMS

Let us establish some notations for use throughout this chapter. Let ξ_n, F_n, and μ_n be respectively, a r.v., its d.f., and the p.m. induced by its d.f., $n \in N^*$. Then we set $Q_n = Q_{\xi_n} = Q_{F_n} = Q_{\mu_n}$, $n \in N^*$. Further, for $m < n$, $m, n \in N$, set

$$\xi_{(m,n)} = \sum_{k=m+1}^{n} \xi_k, \qquad \xi_{(n)} = \xi_{(0,n)}, \qquad \xi_{(0)} = 0;$$

if ξ_n, $n \in N^*$, are i.r.v.'s, then

$$F_{(m,n)} = \mathop{\ast}_{k=m+1}^{n} F_k, \qquad F_{(n)} = F_{(0,n)}, \qquad F_{(0)} = H_0,$$

and

$$\mu_{(m,n)} = \mathop{\ast}_{k=m+1}^{n} \mu_k, \qquad \mu_{(n)} = \mu_{(0,n)}, \qquad \mu_{(0)} = \delta_0.$$

For i.r.v.'s ξ_n, $n \in N^*$, we set $Q_{(m,n)} = Q_{\xi_{(m,n)}} = Q_{F_{(m,n)}} = Q_{\mu_{(m,n)}}$, and $Q_{(n)} = Q_{\xi_{(n)}} = Q_{F_{(n)}} = Q_{\mu_{(n)}}$, $Q_{(0)} = H_0$.

We start with

THEOREM 3.2.1

Suppose that $F_n \in \mathfrak{F}$, $n \in N^*$, and $F_n \xrightarrow{\text{w}} F \in \mathfrak{F}$ as $n \to \infty$. Then $Q_n \xrightarrow{\text{w}} Q_F$ as $n \to \infty$.

Proof

By (3.1.3) it follows that $Q_n \xrightarrow{\text{w}} Q \in \mathfrak{F}$ as $n \to \infty$. Since the weak limit is unique, we conclude that $Q = Q_F$. ◇

The converse of Theorem 3.2.1. is false; take as an example

$$F_n = \begin{cases} H_0 & \text{if } n \text{ is even}, \\ H_1 & \text{if } n \text{ is odd}. \end{cases}$$

COROLLARY 1

Suppose that $F_n \in \mathfrak{F}$, $n \in N^*$, $F_n \xrightarrow{\text{w}} F \in \mathfrak{F}$ as $n \to \infty$, and the greatest jump α_n of F_n verifies the condition $\alpha_n \to 0$ as $n \to \infty$. Then F is continuous.

Proof

By (1.1.3) we have $Q_n(0) = \alpha_n$ for every $n \in N^*$, so that $Q_F(0) = 0$, and by Theorem 1.2.2 it follows that F is continuous. ◇

COROLLARY 2

Suppose that $F_n \in \mathfrak{F}$ is defined by

$$F_n(x) = \begin{cases} 0 & \text{if } \ x \leq x_0^n, \\ i/m_n & \text{if } \ x_{i-1}^n < x \leq x_i^n, \quad 1 \leq i \leq m_n - 1 \\ 1 & \text{if } \ x_{m_n - 1}^n < x, \end{cases}$$

where $x_0^n < x_1^n < \cdots < x_{m_n-1}^n$, $|x_0| \leq a < \infty$, $n \in N^*$. If $(m_n)_{n \in N^*}$ is an increasing sequence of natural numbers, and if $F_n \xrightarrow{\text{w}} F \in \mathfrak{F}$ as $n \to \infty$, then F is continuous.

Proof

It suffices to apply Corollary 1 of Theorem 3.2.1 with $\alpha_n = m_n^{-1}$, $n \in N^*$. ◇

Theorem 3.2.1 may be strengthened as follows:

THEOREM 3.2.2

Suppose that $F_n \in \mathfrak{F}$, $n \in N^*$, and $F_{(n)} \xrightarrow{\text{w}} F \in \mathfrak{F}$ as $n \to \infty$. Then

$$\lim_{n \to \infty} Q_{(n)}(l) = Q_F(l) \tag{3.2.1}$$

for every $l > 0$.

Proof

By Theorem 3.2.1 we know that $Q_{(n)} \xrightarrow{\text{w}} Q_F$. Moreover, for every $\varepsilon > 0$, there is an $n_0 = n_0(\varepsilon) \in N^*$, such that for $n \geq n_0(\varepsilon)$ we have

$$Q_F(l) \leq Q_{(n)}(l) \leq Q_F(l + \varepsilon) + \varepsilon$$

for every $\varepsilon > 0$. Since ε is arbitrary, we conclude that (3.2.1) holds true. ◇

Notes and Comments

Theorem 3.2.1 and its Corollaries are given by Sâmboan and Theodorescu (1968). Theorem 3.2.2 is to be found in Tucker (1963).

3.3 ESSENTIAL CONVERGENCE

3.3.1 We start by examining $Q_{(m,n)}$. Clearly $Q_{(m,n)}$ is nonincreasing with respect to n (cf. Theorem 2.1.1) so that the limit $\lim_{n\to\infty} Q_{(m,n)} = Q_{(m,\infty)}$ exists for every fixed $m \in N$. Analogously, $Q_{(m,\infty)}$ is nondecreasing with respect to m (cf. Theorem 2.1.1) so that the limit $\lim_{m\to\infty} Q_{(m,\infty)} = Q$ exists and Q is nondecreasing, of l, $0 \leq Q|l| \leq 1$.

First we prove an auxilliary result.

LEMMA 3.3.1

There is an integer-valued increasing function $\varphi(m) > m$, $m \in N^*$, such that $\lim_{m\to\infty} Q_{(m,\varphi(m))}(l) = Q(l)$ for every continuity point l of Q.

Proof

Let $(a_k)_{k \in N^*}$ be a denumerable set of points which is dense in $[0, \infty)$ (e.g., the rational numbers in $[0, \infty)$) and let $\varepsilon > 0$ be given. Then there exists $m_k = m_k(\varepsilon, a_k)$ such that $|Q_{(m,\infty)}(a_k) - Q(a_k)| \leq \varepsilon/2$ for every $m \geq m_k$. Next, there exists $n_{k,m} = n_{k,m}(\varepsilon, a_k, m) > m$, $m \geq m_k$, such that $|Q_{(m,n)}(a_k) - Q_{(m,\infty)}(a_k)| \leq \varepsilon/2$ for every $n \geq n_{k,m}$. Therefore $|Q_{(m,n)}(a_k) - Q(a_k)| \leq \varepsilon$ for every $m \geq m_k$ and $n \geq n_{k,m} > m$.

Set now $\varphi(m) = \max_{1 \leq k \leq m} n_{k,m}$. Consider an arbitrary point a of the sequence $(a_k)_{k \in N^*}$. For every given $\varepsilon > 0$, there exists $m_1 = m_1(a)$ such that $|Q_{(m,\varphi(m))}(a) - Q(a)| \leq \varepsilon$ for every $m \geq m_1$. By a well-known result, quoted on p. 82, we conclude that $\lim_{m \to \infty} Q_{(m,\varphi(m))}(l) = Q(l)$ for every continuity point l of Q. ◇

We can now prove

PROPOSITION 3.3.2

The function Q does not depend on l and is either $Q \equiv 0$ or $Q \equiv 1$.

Proof

Put $Q(\infty) = \lim_{l \to \infty} Q(l)$. We distinguish two cases: (a) $Q(\infty) = 0$; (b) $Q(\infty) > 0$.

(a) If $Q(\infty) = 0$, then $Q \equiv 0$ since Q is nondecreasing. This is equivalent to $Q_{(m,\infty)} = 0$ for every $m \in N^*$, so that $Q_{(0,\infty)} \equiv 0$.

(b) If $Q(\infty) > 0$ there exists an $l_0 > 0$ which is a continuity point of Q, such that $Q(l_0) > 0$, and for every $\varepsilon > 0$ there exists $m_0 = m_0(\varepsilon) \in N^*$ such that $Q_{(m,\infty)}(l_0) > 0$ for every $m \geq m_0$.

Set now for a fixed $m \geq m_0$

$$\varphi_k(m) = \begin{cases} \underbrace{\varphi \circ \cdots \circ \varphi}_{k \text{ times}}(m) = s_k & \text{if } k \in N^*, \\ m = s_0 & \text{if } k = 0. \end{cases}$$

Next, let $0 < l \leq l_0$ be an arbitrary continuity point of Q. By Theorem 2.1.1 and (2.2.33) we have

(3.3.1)

$$0 < Q_{(m,\infty)}(l_0) \leq Q_{(m,s_{k+1})}(l_0) = Q_C(l_0) \quad \left(C = \mathop{*}_{j=0}^{k} \mu_{(s_j, s_{j+1})}\right)$$

$$\leq A(l_0/l)\left(\sum_{j=0}^{k} (1 - Q_{(s_j, s_{j+1})}(l))\right)^{-1/2}.$$

Since the left member of (3.3.1) is positive and independent of k and l, it follows that

$$\sum_{j \in N} (1 - Q_{(s_j,\, s_{j+1})}(l)) < \infty; \quad \text{hence} \quad \lim_{j \to \infty} Q_{(s_j,\, s_{j+1})}(l) = 1.$$

On the other hand, $(Q_{(s_j,\, s_{j+1})})_{j \in N}$ is a subsequence of $(Q_{(m,\, \varphi(m))})_{m \in N^*}$ and therefore, by Lemma 3.3.1,

$$\lim_{n \to \infty} Q_{(m,\, \varphi(m))}(l) = Q(l) = 1. \tag{3.3.2}$$

Since (3.3.2) is true for every continuity point l of Q, it follows that $Q \equiv 1$. ◇

3.3.2 We say that a sequence $(\xi_n)_{n \in N^*}$ of r.v.'s is *essentially convergent* in a particular sense if there is a sequence of centering numbers $(a_n)_{n \in N^*}$ such that the sequence $(\xi_n - a_n)_{n \in N^*}$ is convergent in the same sense; otherwise it is *essentially divergent*. Further, a series of r.v.'s is essentially convergent in a particular sense if the sequence of its partial sums is essentially convergent in the same sense.

THEOREM 3.3.3

The series $\sum_{n \in N^*} \xi_n$ of i.r.v.'s is either essentially divergent in law (iff $Q \equiv 0$) or essentially convergent in probability (iff $Q \equiv 1$).

Proof

(a) Let $Q \equiv 0$. Suppose that for each $n \in N^*$ there exists a constant a_n such that $\xi_{(n)} - a_n \xrightarrow{\text{L}} \xi$ as $n \to \infty$. Then by Theorem 3.2.2

$$\lim_{n \to \infty} Q_{\xi(n) - a_n}(l) = \lim_{n \to \infty} Q_{(n)}(l) = Q_\xi(l)$$

for every $l \geq 0$ which contradicts $Q_\xi \equiv 0$.

(b) Let $Q \equiv 1$. There exists a sequence of natural numbers $(m_k)_{k \in N^*}$ such that $m_k \to \infty$ as $k \to \infty$ and $Q_{(m,\, \infty)}(2^{-k}) > 1 - 2^{-k}$ for every $m \geq m_k$. Therefore $Q_{(m,\, n)}(2^{-k}) > 1 - 2^{-k}$ for every $n > m \geq m_k$.

We now construct a_n such that $(\eta_n = \xi_{(n)} - a_n)_{n \in N^*}$ mutually con-converges in probability. Namely, for $m_k < n \leq m_{k+1}$ there is an x_n [cf. (1.1.4)] such that

$$\mu_{(m_k, n)}([x_n, x_n + 2^{-k}]) = Q_{(m_k, n)}(2^{-k}) > 1 - 2^{-k}.$$

Put

$$a_n = \sum_{m_k < n} x_{m_k} + x_n .$$

Then for every $m_k < n \leq m_{k+1}$ we have

$$P(|\eta_n - \eta_{m_k}| > 2^{-k-1}) = 1 - Q_{(m_k, n)}(2^{-k}) < 2^{-k}.$$

Further, let ε and δ be positive. Then there exists $m_{k_0} = N_0(\varepsilon, \delta)$ such that $2^{1-m_{k_0}} < \inf(\varepsilon/3, \delta/3)$. If $(\phi_m)_{m \in N^*}$ is an arbitrary sequence of i.r.v.'s, then there exists an $a > 0$ such that

$$\text{(3.3.3)} \qquad P(|\phi_n - \phi_m| < a) \leq \sum_{k=m+1}^{n} P(|\phi_k - \phi_{k-1}| > a2^{-k}) \leq \sum_{k \in N^*} P(|\phi_k - \phi_{k-1}| > a2^{-k})$$

for every $n > m$.

Let $\bar{n} = \sup_{m_k < n} m_k$, $m_0 = 0$. Then we have for $n > m \geq m_{k_0}$

$$P(|\eta_n - \eta_m| > \delta) \leq P(|\eta_m - \eta_{\bar{m}}| > \delta/3) + P(|\eta_n - \eta_{\bar{n}}| > \delta/3) + P(|\eta_{\bar{n}} - \eta_{\bar{m}}| > \delta/3)$$

if $\bar{n} < \bar{m}$, and

$$P(|\eta_n - \eta_m| > \delta) \leq P(|\eta_m - \eta_{\bar{m}}| > \delta/3) + P(|\eta_n - \eta_{\bar{n}}| > \delta/3)$$

if $\bar{n} = \bar{m}$. On the other hand, since $m, n \geq m_{k_0}$, we have that

$$P(|\eta_n - \eta_{\bar{n}}| > \delta/3) \leq P(|\eta_n - \eta_{\bar{n}}| > 2^{-m_{k_0}}) \leq P(|\eta_n - \eta_{\bar{n}}| > 2^{-\bar{n}-1}) < 2^{-\bar{n}} \leq 2^{-m_{k_0}} < \varepsilon/3$$

and

$$P(|\eta_m - \eta_{\bar{m}}| > \delta/3) < \varepsilon/3.$$

Moreover, if $\bar{n} > \bar{m}$, then by (3.3.3)

$$\begin{aligned} P(|\eta_{\bar{n}} - \eta_{\bar{m}}| > \delta/3) &\leq \sum_{k \geq k_0} P(|\eta_{m_{k+1}} - \eta_{m_k}| > (\delta/3)2^{-k}) \\ &\leq \sum_{k \geq k_0} P(|\eta_{m_{k+1}} - \eta_{m_k}| > 2^{-m_{k_0} - k}) \\ &\leq \sum_{k \in N} P(|\eta_{m_{k_0+k+1}} - \eta_{m_{k_0+k}}| > 2^{-m_{k_0} - k_0 - k}) \\ &\leq 2^{-m_{k_0}} < \varepsilon/3; \end{aligned}$$

this implies that for any positive ε and δ there exists a natural number $N(\varepsilon, \delta)$ such that

$$P(|\eta_n - \eta_m| > \delta) < \varepsilon$$

for every $n > m \geq N(\varepsilon, \delta)$, that is $(\eta_n)_{n \in N^*}$ mutually converges in probability. According to the Cauchy criterion, η_n converges in probability to a limit, say ξ, as $n \to \infty$, and a fortiori converges in law to the same limit. ◇

3.3.3 Suppose that ξ_n, $1 \leq k \leq n$ is a finite family of i.r.v.'s. Then there exist constants a_k, $m \leq k \leq n$, such that $\eta_k = \xi_{(k)} - a_k$, $m \leq k \leq n$, and $\text{med}(\eta_n - \eta_k) = 0$ for every $m \leq k \leq n - 1$. We take for instance $a_n = \text{med}\, \xi_{(n)}$ and $-a_k = \text{med}(\eta_n - \xi_{(k)})$, $m \leq k \leq n - 1$. This property is already verified for $\xi_{(k)}$, $1 \leq k \leq n$ (i.e. $a_k \equiv 0$, $1 \leq k \leq n$), if the ξ_k are all symmetric about the origin.

We can now prove the following auxiliary inequality.

PROPOSITION 3.3.4

Let ξ_k, $1 \leq k \leq n$, be i.r.v.'s and put

$$T = \max_{m \leq k \leq n} |\eta_k - \eta_m|,$$

where $\eta_k = \xi_{(k)} - a_k$, $1 \leq k \leq n$, and $\text{med}(\eta_k - \eta_n) = 0$, $m \leq k \leq n - 1$. Then

$$P(T > 2l) \leq 2(1 - Q_{(m,n)}(l)) \tag{3.3.4}$$

for every $l > 0$.

Proof

Clearly (3.3.4) is satisfied if either $Q_{(m,n)}(l) \le \frac{1}{2}$ or $P(T > 2l) = 0$. Therefore let us suppose that $P(T > 2l) > 0$ and $Q_{(m,n)}(l) > \frac{1}{2}$.

For all $m \le k \le n$, define the events

$$E_k = \{|\eta_i - \eta_m| \le 2l, \quad m \le i \le k-1, \quad |\eta_k - \eta_m| > 2l\}.$$

They constitute a partition of the event $\{T > 2l\}$, so that

$$P(T > 2l) = \sum_{k=m}^{n} P(E_k) = \sum_{\substack{P(E_k)>0 \\ m \le k \le n}} P(E_k). \tag{3.3.5}$$

On the other hand, for each E_k such that $P(E_k) > 0$ we have

$$P(|\eta_n - \eta_m| \le l \mid E_k) \le P(|\eta_n - \eta_k| > l \mid E_k) = P(|\eta_n - \eta_k| > l). \tag{3.3.6}$$

The inequality in (3.3.6) follows because E_k and $\{|\eta_n - \eta_m| \le l\}$ imply $\{|\eta_n - \eta_k| > l\}$. The equality in (3.3.6) from the independence of E_k and $\{|\eta_n - \eta_k| > l\}$.

Further, since $Q_{(m,n)}(l) > \frac{1}{2}$ it follows (cf. Theorem 2.1.1) that $Q_{(k,n)}(l) > \frac{1}{2}$ and so there exists an x_k such that

$$P(|\eta_n - \xi_{(k)} - x_k| \le l/2) = Q_{(k,n)}(l) > \tfrac{1}{2}$$

for every $m \le k \le n-1$. Therefore $\text{med}(\eta_n - \xi_{(k)})$ lies in the interval $[x_k - l/2,\ x_k + l/2]$; hence

$$P(|\eta_n - (\xi_{(k)} - \text{med}(\xi_{(k)} - \eta_n))| \le l) = P(|\eta_n - \eta_k| \le l) \ge Q_{(k,n)}(l)$$

and consequently

$$P(|\eta_n - \eta_k| > l) \le 1 - Q_{(k,n)}(l) \le 1 - Q_{(m,n)}(l) \tag{3.3.7}$$

for every $m \le k \le n-1$. By (3.3.5)–(3.3.7) we can write

$$\begin{aligned} P(|\eta_n - \eta_m| \le l, T > 2l) &= P(|\eta_n - \eta_m| \le l, \bigcup_{\substack{P(E_k)>0 \\ m \le k \le n}} E_k) \\ &= \sum_{\substack{P(E_k)>0 \\ m \le k \le n}} P(|\eta_n - \eta_m| \le l \mid E_k) P(E_k) \\ &\le \sum_{\substack{P(E_k)>0 \\ m \le k \le n}} P(|\eta_n - \eta_k| > l) P(E_k) \\ &\le (1 - Q_{(m,n)}(l)) \sum_{\substack{P(E_k)>0 \\ m \le k \le n}} P(E_k) \\ &= (1 - Q_{(m,n)}(l)) P(T > 2l). \end{aligned} \tag{3.3.8}$$

But

$$(3.3.9) \quad P(|\eta_n - \eta_m| \le l, T \le 2l) \le P(T \le 2l) = 1\, P - (T > 2l),$$

so that from (3.3.8) and (3.3.9) we get

$$(3.3.10) \qquad P(|\eta_n - \eta_m| \le l) \le 1 - Q_{(m,n)}(l) P(T > 2l).$$

By (3.3.7) for $k = m$ we obtain $P(|\eta_n - \eta_m| > l) \le 1 - Q_{(m,n)}(l)$, so that 3.3.10) becomes

$$P(T > 2l) \le \frac{1 - Q_{(m,n)}(l)}{Q_{(m,n)}(l)} \le 2(1 - Q_{(m,n)}(l)),$$

i.e. (3.3.4). ◇

We can now establish the main result

THEOREM 3.3.5

Essentially convergence in law, in probability and a.s. are equivalent for a series $\sum_{n \in N^*} \xi_n$ of i.r.v.'s. In other words, $\sum_{n \in N^*} \xi_n$ is essentially convergent a.s. iff $Q(l) \equiv 1$.

Proof

By Theorem 3.3.3 it suffices to prove that $Q(l) \equiv 1$ implies that $\sum_{n \in N^*} \xi_n$ is essentially convergent a.s. In this case there exists a_n such that $\eta_n = \xi_{(n)} - a_n$, $n \in N^*$, converges in probability to ξ as $n \to \infty$ and therefore there exists a subsequence $(\eta_{n_p})_{p \in N^*}$ which converges a.s. to ξ as $p \to \infty$ such that

$$P(|\eta_{n_{p+1}} - \eta_{n_p}| \ge 2^{-p}) < 2^{-p}$$

for every $p \in N^*$.

Now let us keep fixed the subsequence $(\eta_{n_p})_{p \in N^*}$ and let us modify the elements of the sequence $(\eta_n)_{n \in N^*}$ which do not belong to the mentioned subsequence. Let us set therefore

$$\zeta_q = \begin{cases} \eta_q & \text{if } \; q = n_p, \quad p \in N^*, \\ \eta_q - b_q & \text{if } \; n_p < q < n_{p+1}, \quad p \in N^* \end{cases}$$

such that $\text{med}(\zeta_{n_{p+1}} - \zeta_q) = 0$ for every $n_p \leq q \leq n_{p+1} - 1$ and for every $p \in N^*$. For instance choose first $b_{n_{p+1}-1}$ such that $\text{med}(\zeta_{n_{p+1}-1} - \zeta_{n_p}) = 0$, then take $q, n_p < q < n_{p+1} - 1$ such that $\text{med}(\zeta_{n_{p+1}-1} - \zeta_q) = 0$. Then let us apply Proposition 3.3.4 for the group $\zeta_{n_p}, \ldots, \zeta_{n_{p+1}-1}$. It follows that

$$\begin{aligned} P(\max_{n_p<q<n_{p+1}} |\zeta_q - \zeta_{n_p}| > 2^{-p+1} &= P(\max_{1\leq k<n_{p+1}-n_p} |v_k| > 2^{-p+1}) \\ &\leq 2(1 - Q_{(n_p, n_{p+1})}(2^{-p})) \\ &\leq 2P(|\zeta_{n_{p+1}} - \zeta_{n_p}| > 2^{-p}) \\ &\leq 2^{-p+1}. \end{aligned}$$

Put

$$T_p = \max_{n_p<q<n_{p+1}} |\zeta_q - \zeta_{n_p}|,$$

for every $p \in N^*$. Since

$$P(T_p > 2^{-p+1}) < 2^{-p+1},$$

it follows that T_p converges a.s. to 0 as $p \to \infty$. Therefore,

$$\begin{aligned} |\xi - \zeta_n| &\leq |\zeta_n - \zeta_{n_p}| + |\zeta_{n_p} - \xi| \\ &\leq T_p + |\xi - \xi_{(n_p)} + a_{n_p}| \to 0 \quad \text{a.s.}, \end{aligned}$$

that is, $\zeta_n \to \xi$ a.s. ◇

3.3.4 We now deduce several consequences of the preceding theorems.

THEOREM 3.3.6

$Q_{(n)}$ converges pointwise as $n \to \infty$ to a not identically vanishing function $Q_{(0,\infty)}$ iff $\sum_{n\in N^*} \xi_n$ is a.s. essentially convergent. In this case, $Q_{(0,\infty)}$ is the cn.f. of $\sum_{n\in N^*} (\xi_n - a_n)$, where $(a_n)_{n\in N^*}$ are centering numbers.

Proof

In fact, if there is an $L > 0$ such that $\lim_{n\to\infty} Q_{(n)}(L) = a > 0$, then $Q(L) \geq a > 0$. By Proposition 3.3.2 it follows that $Q \equiv 1$ and by Theorem 3.3.3 that $\sum_{n\in N^*} \xi_n$ is a.s. essentially convergent.

Conversely, if $\sum_{n \in N^*} \xi_n$ is a.s. essentially convergent, then $\sum_{n \in N^*} (\xi_n - a_n)$ converges for some sequence $(a_n)_{n \in N^*}$. Since cn.f.'s remain invariant under shifts, the conclusion follows from Theorem 3.2.2. $\diamondsuit$

From Theorem 3.3.6 we deduce some corollaries.

COROLLARY 1

Suppose that the series $\sum_{n \in N^*} \xi_n$ of i.r.v.'s converges a.s. Then its d.f. is continuous iff $Q_{(n)}(0) \to 0$ as $n \to \infty$.

Proof

This follows from Theorem 3.3.6 and 1.2.2 and Corollary 1 of Theorem 3.2.1. $\diamondsuit$

COROLLARY 2

Suppose that the series $\sum_{n \in N^*} \xi_n$ of discrete i.r.v.'s converges a.s. Then its d.f. is discrete iff $Q_{(n)}(0) \to$ (some) $a > 0$ as $n \to \infty$.

Proof

By Corollary 1 of Theorem 3.3.6 the d.f. of $\sum_{n \in N^*} \xi_n$ is not continuous if $\lim_{n \to \infty} Q_{(n)}(0) > 0$. By Theorem 35† of Jessen and Wintner (1935, p. 86) [see also Lukacs (1970, p. 64)], we conclude that the d.f. of $\sum_{n \in N^*} \xi_n$ is discrete. $\diamondsuit$

† This theorem states that if $(\xi_n)_{n \in N^\circ}$ is a sequence of discrete i.r.v.'s, and if $\sum_{n \in N^*} \xi_n$ converges in law, then this sum is either discrete, or singular or absolutely continuous.

COROLLARY 3

Let $(\xi_n)_{n \in N^*}$ be a sequence of i.r.v.'s and suppose that there is a r.v. ζ such that the r.v.'s η_n defined by $\zeta = \xi_1 + \cdots + \xi_n + \eta_n$ are independent of ξ_k, $1 \leq k \leq n$, for every $n \in N^*$. Then the series $\sum_{n \in N^*} \xi_n$ is a.s. essentially convergent.

Proof

By Theorem 2.1.1, $Q_{(n)}(l) \geq Q_\zeta(l)$ for every $n \in N^*$ and $l \geq 0$, and therefore $\lim_{n \to \infty} Q_n(l) \geq Q_\zeta(l) > 0$ for every $l > 0$. The conclusion follows from Theorem 3.3.6. ◇

COROLLARY 4

Suppose that $0 < a_{n+1} < a_n$, $n \in N^*$, and that $(\xi_n)_{n \in N^*}$ is a sequence of i.r.v.'s such that $P(\xi_n = \pm a_n) = \frac{1}{2}$ for every $n \in N^*$. Then the series $\sum_{n \in N^*} \xi_n$ is a.s. convergent and its d.f. is continuous iff $\sum_{n \in N^*} a_n^2 < \infty$.

Proof

Convergence a.s. is necessary and sufficient for $\sum_{n \in N^*} a_n^2 < \infty$ by Kolmogorov's three series criterion [see e.g., Loève (1963, p. 237)]. By Proposition 2.2.14 we get $Q_{(n)}(0) \to 0$ as $n \to \infty$. Corollary 1 of Theorem 3.3.6 implies then that the d.f. of $\sum_{n \in N^*} \xi_n$ is continuous. ◇

3.3.5 We have seen (c.f. Theorem 3.2.1) that $F_n \xrightarrow{w} F \in \mathfrak{F}$ as $n \to \infty$ implies $Q_n \xrightarrow{w} Q_F$. More generally, $F_{n, a_n} \xrightarrow{w} F \in \mathfrak{F}$ as $n \to \infty$ leads to the same conclusion. In other words weak essential convergence of the F_n implies weak convergence of the Q_n. The converse is false. However, there are two important special cases in which it holds. The first has been stated in Theorem 3.3.6. The second one is given now.

THEOREM 3.3.7

If $Q_n \xrightarrow{w} H_0$ as $n \to \infty$, then μ_n is weakly essentially convergent to δ_0 as $n \to \infty$.

Proof

For every $\varepsilon > 0$ and $l > 0$ choose $n_0 = n_0(l, \varepsilon)$ sufficiently large so that $Q_n(l) > 1 - \varepsilon > \frac{1}{2}$ for every $n \geq n_0$. Now, there is a number $x_n \in R$ such that $\mu_n([0, l] + x_n) \geq Q_n(l) - 1/n$. Let $\mathcal{O}$ be an arbitrary open set. If $0 \notin \mathcal{O}$, then $\liminf_{n\to\infty} \mu_{n, x_n}(\mathcal{O}) \geq 0$; on the other hand, if $0 \in \mathcal{O}$, there exists $[-l, l] \subset \mathcal{O}$, $l > 0$, such that $\mu_{n, x_n}(\mathcal{O}) \geq Q_n(l) - 1/n > 1 - \varepsilon$ for $n \geq n_0$, therefore $\liminf_{n\to\infty} \mu_{n, x_n}(\mathcal{O}) \geq 1 - \varepsilon$ for every $\varepsilon > 0$. Hence $\mu_{n, x_n} \xrightarrow{w} \delta_0$ as $n \to \infty$, i.e., μ_n is weakly essentially convergent to δ_0 as $n \to \infty$. ◇

3.3.6 We now give a result concerning the divergence case.

We say that a numerical sequence $(a_n)_{n \in N^*}$ belongs to the *upper class* or to the *lower class* of a sequence $(\xi_n)_{n \in N^*}$ of r.v.'s, according as $P(\xi_n > a_n \text{ i.o.}) = 0$ or 1.

THEOREM 3.3.8

If the series $\sum_{n \in N^*} \xi_n$ of i.r.v.'s is a.s. essentially divergent, then every numerical sequence $(a_n)_{n \in N^*}$ belongs to either the upper or the lower class of the sequence $(\xi_{(n)})_{n \in N^*}$.

Proof

We shall prove that $P(\xi_{(n)-a_n} > 0 \text{ i.o.}) = P(\limsup_{n\to\infty} \xi_{(n)-a_n} > 0)$ is either 0 or 1.

Set $\alpha_n(x) = P(\limsup_{n\to\infty} \xi_{(n)-a_n} > 0 \mid \xi_{(n)} = x)$; α_n is nondecreasing with respect to x and the limits $\alpha_n(-\infty)$ and $\alpha_n(\infty)$ are well defined. We now show that $\alpha_n(x) = \alpha_n(x+c)$ for any $c \in R$. Indeed, we have

$$\left| P(\limsup_{n\to\infty} \xi_n - a_n + c > 0) - P(\limsup_{n\to\infty} \xi_{(n)} - a_n > 0) \right|$$
$$= \left| \int_R (\alpha_n(x+c) - \alpha_n(x))\, dF_{(n)}(x) \right|$$
$$= \left| \int_R [F_{(n)}(x) - F_{(n)}(x-c)]\, d\alpha_n(x) \right| \leq Q_{(n)}(c).$$

Since $\sum_{n \in N^*} \xi_{(n)}$ as a.s. essentially divergent, by Theorem 3.3.6 $Q_{(n)} \to 0$ as $n \to \infty$; therefore $\alpha_n(x) = \alpha_n(x+c)$ $\mu_{(n)}$-a.s. for any $c \in R$ and by the monotonicity of α_n we conclude that $\alpha_n(x) = \alpha_n(x+c)$ for any $c \in R$. Thus $\alpha_n(x) = \text{const}$.

The assertion of the theorem follows immediately from the zero–one law. ◇

Complements

1 If the series $\sum_{n \in N^*} \xi_n$ is a.s. essentially convergent, then the constants $\bar{a}_n = \text{med } \xi_n + E(\xi_n - \text{med } \xi_n)^c$, where $\xi^c = \xi$ or 0 according as $|\xi| < c$ or $|\xi| \geq 0$, are unconditionally centering [Loève (1963, p. 538)].

2 Conditions for a.s. convergence of sums of i.r.v.'s were obtained by means of centerings at medians. The methods continue to apply in the general case, provided the centering quantities are conditioned and, thus, become themselves r.v.'s. Furthermore, as is expected, the conditions so obtained will be sufficient but no longer necessary. Proofs run parallel to these in the case of independence. We quote one result as an example. Let $(\xi_n)_{n \in N^*}$ be an arbitrary sequence of r.v.'s. If $\xi_{(n)} \xrightarrow{\text{P}} \xi$, then there exists a sequence $(a_n)_{n \in N^*}$ of conditional medians of suitably selected partial sums such that $a_n \xrightarrow{\text{P}} 0$ as $n \to \infty$ and $\xi_{(n)} - a_n \xrightarrow{\text{a.s.}} \xi$ as $n \to \infty$. In other words, $\sum_{n \in N^*}$ is a.s. essentially convergent [Loève (1963, p. 385)].

3 Let $\gamma_k(l) = \mu_k([-e, e]^c)$, $1 \leq k \leq n$, and $\gamma_{(n)}(l) = \sup_{1 \leq k \leq n} \gamma_k(l)$. In certain convergence problems the following quantity is used: $\inf_{l>0}(Q_{\mu(n)}(l) + \gamma_{(n)}(l))$ [Lecam (1965b)].

Notes and Comments

Lemma 3.3.1 and Proposition 3.3.2 are due to Lévy (1937, 1954). Here the original proof of Proposition 3.3.2 was simplified by using Kolmogorov-type inequality (2.2.33). The proof of Theorem 3.3.3 indicates a procedure for determining the centering constants a_n, $n \in N^*$. However, this procedure is not as effective as one would like, so others should be used. We mention two more such procedures; the first one concerns the use of medians and was indicated by Lévy (1937, 1954, p. 135–136), whereas the latter uses the so-called Doob centering numbers [see, e.g., Ito (1960, p. 48–49], Loève (1963, p. 537–538)]. Proposition 3.3.4, due to Lévy (1937, 1954), was proved also in a more general context by Parthasarathy (1967, p. 166–168) Theorem 3.3.5 preserves the form given by Lévy (1937, 1954). Theorem 3.3.6 is essentially due to Ito (1960, p. 46), who used c.f.'s to prove it; here the Tucker (1963) proof is given. The proofs of the corollaries of Theorem 3.3.6 are due to Tucker (1963); some of the results are already known [see, e.g., Loève (1963, p. 538), for Corollary 3]. Theorem 3.3.7 is due to Hengartner and Theodorescu. Theorem 3.3.8 is due to Lévy (1937; 1954, p. 147–149) and is strongly connected with the law of the iterated logarithm [see, e.g., Loève (1963, p. 260–263)].

4 CONCENTRATIONS

This chapter is devoted to other notions related to cn.f.'s. Some of them are used in handling convergence problems (Section 4.1) and are real numbers that are associated with p.m.'s. Others are to be met when studying Lebesgue decomposition of measures (Section 4.2). Finally, there exist information-theoretical concepts (Section 4.3), a special case which leads to a cn.

4.1 CONVERGENCE CONCENTRATIONS

4.1.1 Certain convergence problems may be handled by using real numbers associated with p.m.'s; these numbers are called cn.'s. For these problems they yield much the same properties as the cn.f.'s.

Suppose that λ is a p.m. on $\mathscr{B}$ absolutely continuous on R with respect to the Lebesgue measure m, and such that the Radon-Nykodim derivative $d\lambda/dm > 0$ on $(0, \alpha)$, for some $\alpha > 0$. Let h be a continuous mapping from $[0, 1]$ to $[0, 1]$ satisfying the following conditions:

(h_1) for any sequence of c.f.'s $(\psi_n)_{n\in N^*}$, we have

$$\int_R h(|\psi_n(t)|)\,d\lambda(t) \to 0 \text{ or } 1 \text{ iff } \int_R |\psi_n(t)|\,d\lambda(t) \to 0 \text{ or } 1$$

respectively;

(h_2) $h(0) = 0$, $h(1) = 1$.†

Examples of such functions are $h(x) = x^p$, $p > 0$ and any h such that $x^p \le h(x) \le x^q$, $p, q > 0$.

DEFINITION 4.1.1

The real number

$$q_\mu = \int_R h(|\psi_\mu(t)|)\,d\lambda(t), \tag{4.1.1}$$

where ψ_μ is the c.f. of the p.m. μ, is called the *convergence* cn. of μ (with respect to λ and h).

4.1.2 The following theorem summarizes the main properties of q_μ.

THEOREM 4.1.2

We have

(1) $0 < q_\mu \le 1$;

(2) $q_\mu = 1$ iff $\mu = \delta_{x_0}$;

(3) $q_\mu = q_{\mu_a} = q_{\tilde{\mu}}$;

(4) $q_{\mu_n} \to q_\mu$ as $n \to \infty$ if $\mu_n \xrightarrow{\text{w}} \mu$;

(5) $q_{\mu * \nu} \le \min(q_\mu, q_\nu)$, provided that h is increasing;

(6) $q_{\mu_n} \to 0$ as $n \to \infty$ iff $Q_{\mu_n}(l) \to 0$ as $n \to \infty$ for every $l > 0$;

(7) $q_{\mu_n} \to 1$ as $n \to \infty$ iff μ_n is weakly essentially convergent to δ_0 as $n \to \infty$;

(8) $q_{\mu_{(n)}} \to q > 0$ as $n \to \infty$ iff $\mu_{(n)}$ is weakly essentially convergent.

† This assumption does not lead to any loss of generality.

Proof

(1) There is always a neighborhood of 0 in which $|\psi_\mu(t)| > 1 - \varepsilon$, $\varepsilon > 0$, and a neighborhood of 0 in which $h(|\psi_\mu(t)|) > \frac{1}{2}$; hence $\int_R h(|\psi_\mu(t)|)\, d\lambda(t) > 0$, so that $q_\mu > 0$. Next, it is easily seen that $q_\mu \leq 1$.

(2) This property is mainly based on the fact that $|\psi_\mu(t)| \equiv 1$ m-a.s. in a neighborhood of 0 iff $\mu = \delta_{x_0}$.

(3)–(5) These are straightforward consequences of the elementary properties of c.f.'s.

(6) Suppose that $q_{\mu_n} \to 0$ as $n \to \infty$. According to (2.2.1)

$$Q_{\mu_n}(l) \leq A_2 l \int_{|t| \leq 1/l} |\psi(t)|\, dt.$$

Now let l_0 be large enough so that $1/l_0 < \alpha$; then we have

$$Q_{\mu_n}(l_0) \leq A_2 l_0 \int_{|t| \leq \alpha} |\psi_{\mu_n}(t)|\, dt. \tag{4.1.2}$$

But we know that $\int_R h(|\psi_{\mu_n}(t)|)\, d\lambda(t) \to 0$ as $n \to \infty$, and by (h_1) this is true iff $\int_R |\psi_{\mu_n}(t)|\, d\lambda(t) \to 0$ as $n \to \infty$. Moreover, $\int_0^\alpha |\psi_{\mu_n}(t)|\, dt \to 0$ as $n \to \infty$. Indeed, let g denote $d\lambda/dm$. Then $\int_0^\alpha |\psi_{\mu_n}(t)|\, g(t)\, dt \to 0$ as $n \to \infty$. Set $E_k = \{t \in [0, \alpha) : g(t) > 1/k\}$; then $\bigcup_{m \in N^*} E_m = [0, \alpha)$. Clearly, $\int_{E_k} |\psi_{\mu_n}(t)|\, dt \leq k \int_{E_k} |\psi_{\mu_n}(t)|\, g(t)\, dt \to 0$ an $n \to \infty$. Next, $\varepsilon > 0$ being given, let k_0 be chosen such that $m([0, \alpha) - E_k) < \varepsilon$ for $k > k_0$; then $\int_0^\alpha |\psi_{\mu_n}(t)|\, dt \leq \int_{E_k} |\psi_{\mu_n}(t)|\, dt + \varepsilon$. Hence $\limsup_{n \to \infty} \int_0^\alpha |\psi_{\mu_n}(t)|\, dt \leq \varepsilon$ for every $\varepsilon > 0$. Since $|\psi_{\mu_n}(-t)| = |\psi_{\mu_n}(t)|$ for every $t \in R$, we conclude that $\int_{|t| \leq \alpha} |\psi_{\mu_n}(t)|\, dt \to 0$ as $n \to \infty$. Hence by (4.1.2) we obtain $Q_{\mu_n}(l_0) \to 0$ as $n \to \infty$; clearly, it follows that $Q_{\mu_n}(l) \to 0$ as $n \to \infty$ for every $l > 0$.

Conversely, suppose that for every $l > 0$, $Q_{\mu_n}(l) \to 0$ an $n \to \infty$. Since λ is absolutely continuous with respect to m we have $\psi_\lambda(t) \to 0$ as $|t| \to \infty$ [cf. Lukacs (1970, p. 19)]. For any $\varepsilon > 0$, there is a number $t_0 > 0$ such that $|\psi_\lambda(t)| < \varepsilon$ if $|t| \geq t_0$. Now $q_{\mu_n} \to 0$ as $n \to \infty$ iff $\int_R |\psi_{\mu_n}(t)|^2 \, d\lambda(t) \to 0$ as $n \to \infty$. But

$$\begin{aligned}
\int_R |\psi_{\mu_n}(t)|^2 \, d\lambda(t) &= \iint_{R \times R} \psi_\lambda(x-y) \, d\mu_n(x) \, d\mu_n(y) \\
&\leq \iint_{|x-y| \leq t_0} d\mu_n(x) \, d\mu_n(y) \\
&\quad + \iint_{|x-y| > t_0} |\psi_\lambda(x-y)| \, d\mu_n(x) \, d\mu_n(y) \\
&\leq Q_{\mu_n}(2t_0) + \varepsilon.
\end{aligned}$$

Therefore $\limsup_{n \to \infty} q_{\mu_n} \leq \varepsilon$ for every $\varepsilon > 0$; hence $q_{\mu_n} \to 0$ as $n \to \infty$.

(7) Suppose that $q_{\mu_n} \to 1$ as $n \to \infty$. Then $\int_R |\psi_{\mu_n}(t)| \, d\lambda(t) \to 1$ as $n \to \infty$; since λ is absolutely continuous on R with respect to m and $d\lambda/dm > 0$ in $(0, \alpha)$, we have $\int_0^\alpha (1 - |\psi_{\mu_n}(t)|) \, dt \to 0$ as $n \to \infty$, and

$$\int_0^\alpha (1 - |\psi_{\mu_n}(t)|^2) \, dt \leq 2 \int_0^\alpha (1 - |\psi_{\mu_n}(t)|) \, dt \to 0$$

as $n \to \infty$. Let us show now that $\int_0^{2\alpha} (1 - |\psi_{\mu_n}(t)|^2) \, dt \to 0$ as $n \to \infty$. Indeed, by a simple property of the c.f.'s [see e.g., Lukacs (1970, p. 56)] we have

$$\begin{aligned}
\int_0^{2\alpha} (1 - |\psi_{\mu_n}(t)|^2) \, dt &= 2 \int_0^\alpha (1 - |\psi_{\mu_n}(2t)|^2) \, dt \\
&\leq 8 \int_0^\alpha (1 - |\psi_{\mu_n}(t)|^2) \, dt \to 0
\end{aligned}$$

as $n \to \infty$. Continuing this procedure, we conclude that

$\int_0^{2^k\alpha} (1 - |\psi_{\mu_n}(t)|^2)\,dt \to 0$ as $n \to \infty$ for every $k \in N^*$. Since $1 - |\psi_{\mu_n}(t)|^2 \geq 0$ for all $t \in R$, we get

$$\int_0^T (1 - |\psi_{\mu_n}(t)|^2)\,dt \to 0 \tag{4.1.3}$$

as $n \to \infty$ for all $T > 0$.

Let us prove now that the d.f. $F_n^s = F_n * \tilde{F}_n$, whose c.f. is $|\psi_{\mu_n}(t)|^2$, is weakly convergent to H_0 as $n \to \infty$. Denote by G_n the convolution of F_n^s with the normal d.f. $\Phi(0, \sigma^2; \cdot)$, $n \in N^*$, with a fixed $\sigma > 0$. Then G_n has the c.f. $|\psi_{\mu_n}(t)|^2 \exp(-\frac{1}{2}\sigma^2 t^2)$. Moreover, G_n is symmetric, so that $G_n(0) = \frac{1}{2}$. If we use the classical inversion formula [see, e.g. Lukacs (1970, p. 31)] then

$$\begin{aligned} G_n(x) - \tfrac{1}{2} &= \pi^{-1} \int_0^\infty (\sin tx/t)|\psi_{\mu_n}(t)|^2 \exp(-\tfrac{1}{2}\sigma^2 t^2)\,dt \\ &= \pi^{-1} \int_0^\infty (\sin tx/t) \exp(-\tfrac{1}{2}\sigma^2 t^2)\,dt \\ &\quad + \pi^{-1} \int_0^\infty (\sin tx/t)(|\psi_{\mu_n}(t)|^2 - 1) \exp(-\tfrac{1}{2}\sigma^2 t^2)\,dt \\ &= \Phi(0, \sigma^2; x) - \tfrac{1}{2} + I_{n,\sigma}(x), \end{aligned} \tag{4.1.4}$$

where

$$I_{n,\sigma}(x) = \pi^{-1} \int_0^\infty (\text{xin } tx/t)(|\psi_{\mu_n}(t)|^2 - 1) \exp(-\tfrac{1}{2}\sigma^2 t^2)\,dt.$$

Let us evaluate $I_{n,\sigma}$. $\varepsilon > 0$ being given, choose $T_0 = T_0(\sigma, \varepsilon)$ such that $\int_{T_0}^\infty \exp(-\frac{1}{2}\sigma^2 t^2)\,dt < \varepsilon/2$. Moreover, in virtue of (4.1.3), for $n \geq n_0(\sigma, \varepsilon, x)$, we have

$$\left| \int_0^{T_0} (\sin tx/t)(|\psi_{\mu_n}(t)|^2 - 1) \exp(-\tfrac{1}{2}\sigma^2 t^2)\,dt \right| \leq |x| \int_0^{T_0} (1 - |\psi_{\mu_n}(t)|^2)\,dt < \varepsilon/2.$$

Thus $|I_{n,\sigma}| < \varepsilon$ and by (4.1.4), $G_n \to \Phi(0, \sigma^2; \cdot)$ as $n \to \infty$ uniformly in any finite interval. Consequently $F_n^s \xrightarrow{w} H_0$ as $n \to \infty$ and by

Theorem 3.2.1 $Q_{F_n^s} \xrightarrow{w} H_0$ as $n \to \infty$. Since $Q_{\mu_n} = Q_{F_n} \geq Q_{F_n^s}$, we conclude that $Q_{\mu_n} \xrightarrow{w} H_0$ as $n \to \infty$, and by Theorem 3.3.7 it follows that μ_n is weakly essentially convergent to δ_0 as $n \to \infty$.

Conversely, suppose that μ_n is weakly essentially convergent to δ_0 as $n \to \infty$. In other words, there exist constants a_n, $n \in N^*$, such that $\psi_{\mu_{n,a_n}}(t) \to 1$ as $n \to \infty$ for every $t \in R$. Therefore $\int_R h(|\psi_{\mu_{n,a_n}}(t)|)\, d\lambda(t) \to 1$ as $n \to \infty$, so that it follows from (3) that $q_{\mu_n} \to 1$ as $n \to \infty$.

(8) By (5) we have $q_{\mu_{(n)}} \to 0$ as $n \to \infty$ iff $Q_{\mu_{(n)}}(l) \to 0$ as $n \to \infty$ for every $l > 0$, and this is true iff $\mu_{(n)}$ is weakly essentially divergent. ◇

We note that an important special case of convergence cn. may be obtained by taking $h(x) = x^2$, $x \in [0, 1]$, and $d\lambda(t) = dt/(1 + t^2)$, $t \in R$.

Finally, we remark that Theorem 4.1.2 also holds in r-dimensional Euclidean space.

4.1.3 We shall now show how q_μ provides information about the discrete part of μ. With this intention, we put $h(x) = x^2$, $x \in [0, 1]$ in (4.1.1). Next, consider a family of p.m.'s λ_τ, $\tau \in (0, \infty)$, having the properties mentioned on p. 101, such that for every $t \in R$, $\psi_{\lambda_\tau}(t) \to \psi(t)$ as $\tau \to 0$, where $\psi(t) = 1$ if $t = 0$ and $\psi(t) = 0$ if $t \neq 0$. We now give an example. Suppose that the p.m. λ_τ, $\tau \in (0, \infty)$, is defined by its c.f. $\psi_{\lambda_\tau}(t) = \psi_\lambda(t/\tau)$, where λ is a p.m. with the properties described on p. 101. By Lukacs (1970, p. 19), we conclude that for these ψ_{λ_τ}, we have for any $t \in R$, $\psi_{\lambda_\tau}(t) \to \psi(t)$ as $\tau \to 0$. As a special case, take $\psi_\lambda(t) = \exp(-|t|)$, $t \in R$. If we suppose that λ has a finite variance $\sigma_\lambda^2 > 0$, then we may take $\tau = \sigma_\lambda/\sigma_{\lambda_\tau}$.

For every $\tau \in R$, set now

$$q_\mu = q_\mu(\tau) = \begin{cases} 0 & \text{if } \tau \leq 0 \\ \int_R |\psi_\mu(t)|^2\, d\lambda_\tau(t) & \text{if } \tau > 0 \end{cases}$$

THEOREM 4.1.3

We have

$$\lim_{\tau \downarrow 0} q_\mu(\tau) = \sum_{x \in R} \mu^2(\{x\}), \tag{4.1.5}$$

where the summation on the right is to be taken over all mass points (atoms) of μ.

Proof

For every $\tau \in (0, \infty)$ we have

$$q_\mu(\tau) = \int_R |\psi_\mu(t)|^2 \, d\lambda_\tau(t) = \int_R \psi_{\lambda_\tau}(t) \, d\mu^s(t)$$

and letting $\tau \downarrow 0$ we get

$$\lim_{\tau \downarrow 0} q_\mu(\tau) = \int_R \psi(t) \, d\mu^s(t) = \iint_{R^2} \psi(x - y) \, d\mu(x) \, d\tilde{\mu}(y) = \sum_{x \in R} \mu^2(\{x\}),$$

i.e. (4.1.5). ◇

We now indicate two special cases. The first is obtained by means of the uniform p.m., i.e., $d\lambda(t) = dt/2T$ if $|t| \leq T$ and $d\lambda(t) = 0$ if $|t| > T$; here $\tau = 1/T$ and Theorem 4.1.3 reduces to Theorem 3.3.4, Lukacs (1970, p. 42). The latter is obtained by means of the standard normal p.m. whose c.f. is $\psi_\lambda(t) = \exp(-t^2/2)$, $t \in R$; here $\tau = 1/\sigma_{\lambda_\tau}$.

From Theorem 4.1.3 we get immediately the following

COROLLARY

Suppose that

(i) ψ_{λ_τ} is real valued†;

† This is not an essential restriction because we can always go over to λ_τ^s.

(ii) $\psi_{\lambda_\tau}(t)$ is nondecreasing with respect to τ for every $t \in R$;

(iii) $\psi_{\lambda_\tau}(t) \to 1$ as $\tau \to \infty$ for every $t \in R$. Then q_μ^+ is a d.f.

As examples take $\psi_\lambda(t) = \exp(-|t|)$, $t \in R$, or $\psi_\lambda(t) = \exp(-t^2/2)$, $t \in R$, and take $\psi_{\lambda_\tau}(t) = \psi_\lambda(t/\tau)$, $t \in R$, $\tau \in (0, \infty)$.

Complements

1 Let us consider the quantity

$$\hat{\psi}_\mu(\tau) = \int_0^\tau \psi_\mu(t)\,dt = \int_R (\exp i\tau x - 1)/ix\, d\mu(x),$$

i.e., the *integral c.f.* [see, e.g., Loève (1963, p. 189)]. This integral is strongly related to convergence cn.'s defined by (4.1.1). In fact, if μ is symmetric, then for every $\tau > 0$, $\tau^{-1}\hat{\psi}_\mu(\tau) = q_\mu$, when $h(x) = x^2$, $x \in [0, 1]$, and, $d\lambda(t) = dt/\tau$ of $0 \leq t \leq \tau$ and $d\lambda(t) = 0$ otherwise. Moreover, $\tau^{-1}\hat{\psi}_\mu(\tau)$ is a c.f.

2 Let λ be a p.m. with the properties mentioned on p. 101 and set

$$\tilde{q}_\mu = \int_R |\operatorname{Re} \psi_\mu(t)|\, d\lambda(t).$$

Then properties (1)–(4), (6), and (7) (for weak convergence even) of Theorem 4.1.2 hold true.

3 Theorem 4.1.2 holds true for the quantity $\int_R Q_\mu(l)\, d\lambda(l)$, where λ is the same as on p. 101.

4 Let us replace the p.m. λ in Definition 4.1.1 by an arbitrary nonnegative σ-finite measure on $\mathscr{B}$. In this case we must assume that the integral appearing in (4.1.1) exists and is finite. Theorem 4.1.2 may be adapted under appropriate additional conditions. For instance, the first half of (1), (3), (5), and (8) remain valid. As an example, take as λ Lebesgue measure and $h(x) = x^2$; then (4.1.1) becomes

$$q_\mu = \int_R |\psi_\mu(t)|^2 \, dt$$

and according to Pancherel's theorem [see, e.g., Lukacs (1970, p. 76)] we get

$$q_\mu = \int_R f^2(x) \, dx, \tag{4.1.6}$$

where $f = d\mu/dm$.

5 Suppose that we take in (4.1.1) $h(x) = -\log x$, $x \in [0, 1]$, and $d\lambda(t) = dt/\tau$ if $0 \leq \tau \leq \varepsilon$ and $d\lambda(t) = 0$ otherwise. This h does not satisfy all the conditions given on pp. 101–102. The corresponding q_μ, i.e.,

$$q_\mu = -\int_0^\tau \log |\psi_\mu(t)| \, dt, \tag{4.1.7}$$

has meaning only for those μ whose c.f.'s are different from 0 m-a.s. on $(0, \tau)$. This explains why this q_μ yields only a part of the results listed in Theorem 4.1.2 [see, e.g., Lukacs (1970, p. 167–169), and Linnik (1962, p. 54)].

6 The quantity $-\log q_\mu$, where q_μ is given in (4.1.1), is called *degree of d.s. of* μ; for $h(x) = x^2$, $x \in [0, 1]$, and $d\lambda(t) = dt/(1 + t^2)$, $t \in R$, it was considered by Ito (1960, p. 43, formula (11.1)).

Notes and Comments

Definition 4.1.1 and Theorem 4.1.2 are due to Hengartner and Theodorescu (1972b). They represent generalizations of the results of Ito (1960, pp. 42–49), who examined the special case described at the end of Section 4.1.2. Theorem 4.1.3 and its corollary are due to Hengartner and Theodorescu (1972b).

The quantity (4.1.6) is to be found in Grenander (1963, p. 75), and is called (*measure of*) *cn.*; it is introduced with the aim of flattening out distributions over a locally compact commutative group via convolutions. However, the term of *informational energy* given by Onicescu (1966) [see also Pérez (1967b, p. 1342, formula (1.4))] seems more appropriate. For further details on other information-theoretical concepts, see Section 4.3. The quantity (4.1.7) was considered by Hinčin (1937), and according to Linnik (1962, p. 54), is known as the Hinčin functional; in a more general context, it was used by Grenander (1960, p. 77) as *measure of ds.*

4.2 DECOMPOSITION CONCENTRATIONS

4.2.1 We have seen that the Lévy cn.f. Q_μ provides information about the biggest mass point of μ. In other words, it provides a partial information about the discrete part of μ. Moreover, from Theorem 4.1.3. we conclude that q_μ, as a function of τ, provides also certain information about the discrete part of μ.

Now, let us introduce a quantity characterizing the other components of the Lebesgue decomposition of a p.m. With this intention, let μ be a p.m. on $\mathscr{B}$, and let us define on R the following real-valued function

$$\text{(4.2.1)} \qquad \hat{Q}_\mu(l) = \begin{cases} 0 & \text{if} \quad l < 0, \\ \sup\{\mu(A) : m(A) \leq l, \quad A \in \mathscr{B}\} & \text{if} \quad l \geq 0, \end{cases}$$

where m is Lebesgue measure; clearly, for $l \geq 0$ we have

$$\hat{Q}_\mu(l) = \sup\{\mu(A) : m(A) = l, \quad A \in \mathscr{B}\}.$$

DEFINITION 4.2.1

The function $\hat{Q}_\mu$ is called a *decomposition cn. f.*

Obviously $Q_\mu \geq \hat{Q}_\mu$; examples show that this inequality may be strict. Moreover, it is easily shown that

$$\hat{Q}_\mu(l) = \sup\{\mu(A): m(A) = l, \quad A \in \mathscr{J}\}$$

where $\mathscr{J}$ is the family of all finite unions of intervals of R.

4.2.2 The following theorem summarizes the main properties of $\hat{Q}_\mu$.

THEOREM 4.2.2

We have

(1) $\hat{Q}_\mu$ is a d.f., absolutely continuous and concave on $(0, \infty)$;
(2) $\hat{Q}_\mu = \hat{Q}_{\mu_a} = \hat{Q}_{\tilde{\mu}}$;
(3) $\hat{Q}_\mu(l) \equiv \mu([0, l])$ iff $F_\mu \in \mathfrak{F}_+$ and F_μ is concave (here F_μ is the d.f. induced by μ);
(4) $\hat{Q}_\mu(0) = \mu''(R)$, where μ'' is the singular part of μ with respect to m;
(5) $\hat{Q}_{\mu * \nu} \le \min(\hat{Q}_\mu, \hat{Q}_\nu)$.

Proof

(1) It is easily seen that $\hat{Q}_\mu$ is a d.f.

Next, for any $\varepsilon > 0$ there is an $A_i \in B$ such that $m(A_i) \le l_i$ and $\mu(A_i) > \hat{Q}_\mu(l_i) - \varepsilon$, $i = 1, 2$. But $l_1 + l_2 \ge 2m(A_1 \cap A_2) + m(A_1 \cap A_2^c) + m(A_2 \cap A_1^c)$. Divide $A_1 \cap A_2^c$ [respectively $A_2 \cap A_1^c$] into two Borel sets A'_{12} and A''_{12} [respectively A'_{21} and A''_{21}] such that $m(A'_{12}) = m(A''_{12})$ [respectively $m(A'_{21}) = m(A''_{21})$]. Now set $\tilde{A}_1 = A'_{12}$ if $\mu(A'_{12}) \ge \mu(A''_{12})$ or $\tilde{A}_1 = A''_{12}$ if $\mu(A'_{12}) < \mu(A''_{12})$ and $\tilde{A}_2 = A'_{21}$ if $\mu(A'_{21}) \ge \mu(A''_{21})$, or $\tilde{A}_2 = A''_{21}$ if $\mu(A'_{21}) < \mu(A''_{21})$. Then for $E = \tilde{A}_1 \cup \tilde{A}_2 \cup (A_1 \cap A_2)$ we have $\frac{1}{2}(l_1 + l_2) \ge m(E)$ and

$$\hat{Q}_\mu(\tfrac{1}{2}(l_1 + l_2)) \ge \mu(E) \ge \tfrac{1}{2}(\mu(A_1) + \mu(A_2)) \ge \tfrac{1}{2}(\hat{Q}_\mu(l_1) + \hat{Q}_\mu(l_2)) - \varepsilon$$

for any $\varepsilon > 0$; hence $\hat{Q}_\mu$ is concave.

Since $\hat{Q}_\mu$ is concave we have for all $a > 1$ and $l > 0$ $\hat{Q}_\mu(al) \le a\hat{Q}_\mu(l)$. Therefore it follows from Theorem 1.5.6 that $\hat{Q}_\mu$ is absolutely continuous on $(0, \infty)$.

(2) This property is a simple consequence of the fact that $m = m_a = \tilde{m}$.

(3) Take μ such that $F_\mu \in \mathfrak{F}_+$ and is concave. Clearly $\hat{Q}_\mu(l) = \mu([0, l])$ for $l \geq 0$.

(4) Let μ' and μ'' be respectively the absolute continuous and the singular part of μ with respect to m. Then there is $A \in \mathscr{B}$ such that $m(A) = 0$ and $\mu''(A) = \mu''(R)$. Therefore $\hat{Q}_\mu(0) \geq \mu''(R)$. On the other hand, $\mu'(E) = 0$ for all $E \in \mathscr{B}$ with $m(F) = 0$. Hence $\hat{Q}_\mu(0) = \mu''(R)$.

(5) This property follows in the same manner as for Q_μ. ◇

COROLLARY 1

The following three statements are equivalent:

(i) $\hat{Q}_\mu(0) = 0$;

(ii) $\hat{Q}_\mu$ is absolutely continuous on $[0, \infty)$;

(iii) μ is absolutely continuous.

Proof

Suppose (i) valid. Then by Theorem 4.2.2 (1) and (4), (ii) follows. Conversely, (ii) implies (i).

Clearly (i) and (iii) are equivalent by Theorem 4.2.2 (4). ◇

COROLLARY 2

$\hat{Q}_\mu = H_0$ iff the absolute continuous part μ' of μ vanishes.

Proof

It suffices to take into account Theorem 4.2.2 (4). ◇

4.2.3 It should be noted that $\hat{Q}_\mu$ is not at all appropriate for the study of convergence problems. For instance, if $\mu = \frac{1}{2}(\delta_0 + \delta_1)$, then $\mu_{(n)}$ is weakly essentially divergent but $\hat{Q}_{\mu(n)} = H_0$ for all $n \in N^*$. Nevertheless, as we have seen, this function is very useful in characterizing the Lebesgue decomposition of a p.m.

Complements

1 Let $\{\Omega, \mathscr{K}\ \mu, \nu\}$ be a bimeasure space (i.e., a measurable space endowed with two measures) and let us assume that μ and ν are non-negative σ-finite measures. Given $\mathscr{A} \subset \mathscr{K}$, we define a mapping $\hat{Q}_{\mu,\nu;\mathscr{A}}$ from $[0, \tilde{q}(\Omega)]$ into $[0, \mu(\Omega)]$ by

$$\hat{Q}_{\mu,\nu;\mathscr{A}}(l) = \sup\{\mu(A) : \nu(A) \leq l, \quad A \in \mathscr{A}\}, \tag{4.2.2}$$

called *the cn.f. on* $\mathscr{A}$ *of* μ *with respect to* ν; if $\mathscr{A} = \mathscr{K}$, for the sake of simplicity we shall write $\hat{Q}_{\mu,\nu}$ for $\hat{Q}_{\mu,\nu;\mathscr{K}}$. Special cases: $\hat{Q}_\mu = \hat{Q}_{\mu,m;\mathscr{I}}$, $\hat{Q}_\mu = \hat{Q}_{\mu,m}$ [c.f (4.2.1)].

Moreover, if $\mathscr{A} \subset \mathscr{A}'$, then $\hat{Q}_{\mu,\nu;\mathscr{A}} \leq \hat{Q}_{\mu,\nu;\mathscr{A}}$ [Raoult (1969)].

2 The upper bound in (4.2.2) is reached; the proof is based on a slight extension of the Neyman–Pearson lemma [Raoult (1969)].

3 If ν is atomless, and if $\mathscr{A}$ is an algebra generating $\mathscr{K}$, then $\hat{Q}_{\mu,\nu;\mathscr{A}} = \hat{Q}_{\mu,\nu}$ [Raoult (1969)]. Special case: $\hat{Q}_{\mu,m;\mathscr{I}} = \hat{Q}_{\mu,m}$ (see p. 111).

4 An extension of (4.2.2) is used in order to give a necessary and sufficient condition for the existence of a Lebesgue decomposition of $(\mu_n)_{n \in N^*}$ with respect to $(\nu_n)_{n \in N^*}$, where $(\{\Omega_n, \mathscr{K}_n, \mu_n, \nu_n\})_{n \in N^*}$ is a sequence of bimeasure spaces [Raoult, (1969, 1970)].

Notes and Comments

The results of this section are essentially based on Raoult's (1969, 1970) papers; here, only a special case is dealt with in more detail.

Different generalizations and properties are indicated in the Complements only because they are beyond the scope of the present monograph. However, the idea of replacing segments of length l by sets of Lebesgue measure l in the definition of the Lévy cn.f. goes back to Lévy (1937, 1954) [see also Loève (1963, p. 265)].

4.3 INFORMATION-THEORETICAL CONCEPTS

4.3.1 Let φ be a continuous convex real-valued function defined on $(0, \infty)$. Expressions such as $\varphi(0)$, $0 \cdot \varphi(0/0)$, and $0 \cdot \varphi(a/0)$, $a \in (0, \infty)$ are to be understood as $\lim_{x \to 0} \varphi(x)$, 0, $\lim_{\varepsilon \downarrow 0} \varepsilon\varphi(a/\varepsilon) = a \lim_{u \to \infty} \varphi(u)/u$, respectively. Obviously $\varphi(0)$ and $0 \cdot \varphi(a/0)$ could be equal to $+\infty$; $-\infty$ is however eliminated by convexity of φ.

Suppose that $\mu \ll \lambda$, $\nu \ll \lambda$, where μ and ν are p.m.'s and λ is a σ-finite measure on the measurable space (*input*) $\{X, \mathscr{X}\}$; we may take, e.g., $\lambda = \mu + \nu$.

DEFINITION 4.3.1

The real number

$$I_\varphi(\mu, \nu) = \int_X v(x)\varphi(u(x)/v(x))d\lambda(x), \tag{4.3.1}$$

where $u = d\mu/d\lambda$ and $v = d\nu/d\lambda$ are finite and nonnegative, is called the *φ-deviation of μ and ν.*

Note that $I_\varphi(\mu, \nu)$ does not depend on λ.

Consider now another measurable space (*output*) $\{Y, \mathscr{Y}\}$ and assume that a transition probability function (*noise*) $\gamma(A \mid x)$, $x \in X$, $A \in \mathscr{Y}$, is given. Set

$$\bar{\mu}(A) = \int_X \lambda(A \mid x)\, d\mu(x), \qquad \bar{\nu}(A) = \int_X \gamma(A \mid x)\, d\nu(x);$$

$\bar{\mu}$ and $\bar{\nu}$ are p.m.'s on $\mathscr{Y}$; and

$$I_\varphi(\mu, \nu) \geq I_\varphi(\bar{\mu}, \bar{\nu}) \tag{4.3.2}$$

[cf. Csiszár (1963, Theorem 1)]; moreover, $I_\varphi(\mu, \nu) \geq \varphi(1) = I_\varphi(\mu, \mu)$. In information-theoretical terms, (4.3.2) says that information is lost when transmitting signals through a channel.

4.3.2 Let us indicate several special cases. We start with $\varphi(x) = x \ln x$; in this case we get *the relative information of* μ *and* ν, i.e.

$$I_{x \ln x}(\mu, \nu) = \int_X u(x) \ln(u(x)/v(x))\, d\lambda(x)$$

$$= \begin{cases} \int_X d\mu/d\nu)(x) \ln(d\mu/d\nu)(x)\, d\nu(x) \\ \qquad = \int_X \ln(d\mu/d\nu)(x)\, d\mu(x) & \text{if} \quad \mu \ll \nu \\ \infty \qquad \text{otherwise.} \end{cases} \tag{4.3.3}$$

The relative information of order α *of* μ *and* ν, $I_\varphi(\mu \| \nu)$, is obtained from $I_\varphi(\mu, \nu)$ if we take $\varphi(x) = -x^\alpha$ if $0 < \alpha < 1$ and $\varphi(x) = x^\alpha$ if $\alpha > 1$. Then

$$I_\alpha(\mu \| \nu) = (1 - \alpha)^{-1} \ln |I_\alpha(\mu, \nu)|, \tag{4.3.4}$$

where $I_\alpha(\mu, \nu)$, *the generalized entropy of order* α *of* μ *and* ν. is $I_\varphi(\mu, \nu)$ for this special φ. Hence $I_\alpha(\mu \| \nu)$ is a nondecreasing function of $I_\alpha(\mu, \nu)$. Moreover, $I_{x \ln x}(\mu, \nu) = \lim_{\alpha \to 1} I_\alpha(\mu \| \nu)$, so that $I_{x \ln x}(\mu, \nu)$ may be called *the relative information of order* 1 *of* μ *and* ν and denote it by $I_1(\mu \| \nu]$.

Take now $\alpha = 2$, $X = R$, and $\nu = m$; suppose also that $\mu \ll m$. Then $I_{x^2}(\mu, m)$ reduces to the quantity (4.1.6), i.e., to a cn.

The total variation of $\mu - \nu$ may also be expressed as a special φ-deviation. Namely, take $\varphi(x) = |x - 1|$.

4.3.3 For the sake of simplicity, take $\{X, \mathscr{X}\} = \{R, \mathscr{B}\}$. We can define the following function on $(0, \infty)$:

$$I_\varphi(\mu, \nu; l) = \sup_{x \in R} \int_x^{x+l} v(x)\varphi(u(x)/v(x))\, d\lambda(x); \tag{4.3.5}$$

If the quantity under the integral sign is nonnegative, then $I_\varphi(\mu, v, \infty) = I_\varphi(\mu, v)$ and this function may be extended to R by setting 0 for $l < 0$, and roughly speaking, it may be considered as a "cn.f."

Complements

1 Using $I_\varphi(\mu, v)$, basically the inequality (4.3.2), a pure information-theoretical proof for the ergodic theorem for homogeneous Markov chains can be given [Csiszár, (1963); see also Rényi (1961) for such an "a.s." information-theoretical proof].

2 For information-theoretical estimators of the average and Bayes risk change in certain statistical decision problems, see Pérez (1967a,b).

Notes and Comments

The φ-deviation of μ and v, $I_\varphi(\mu, v)$, given by (4.3.1) was defined by Csiszár (1963). The relative information (of order 1) of μ and v, $I(\mu \| v)$. given by (4.3.3), was used by several authors under different names: *information for discrimination* [Kullback and Leibler (1951)], *I-divergence* [Rényi (1961)], *gain of information* [Pérez, (1957)], *generalized entropy* [Pinsker (1960)]. The relative information of order α of μ and v, $I_\alpha(\mu \| v)$, given by (4.3.4), as well as the generalized entropy of μ and v, $I_\alpha(\mu, v)$, are to be found in Rényi (1961) [see also Csiszár (1962)]. The quantity $I_{\chi^2}(\mu, m)$, identical to (4.1.6), bridges between information-theoretical concepts and those described in Section 4.1 [see also Sâmboan and Theodorescu (1968)].

5 GENERALIZATIONS

This chapter aims to show how the concept of cn.f. may be extended in more general spaces. Section 5.1 is concerned with preliminary results concerning cn.f.'s in arbitrary spaces. Section 5.2 deals with cn.f.'s in metric spaces, whereas Section 5.3 discusses cn.f.'s in Hilbert spaces.

Throughout this chapter we shall use the same notational conventions used in the preceding chapters.

5.1 PRELIMINARY RESULTS

5.1.1 Let $(X, \mathscr{X})$ be an arbitrary measurable space, let $\mathscr{A} \subset \mathscr{X}$, let T be a mapping from $X \times \mathscr{A}$ into $\mathscr{X}$ so that $T(x; \mathrm{A}) \ni x$, and let

$$\mathscr{T} = \{T(x; A) : x \in X, A \in \mathscr{A}\}.$$

Next, let μ be a p.m. on $\mathscr{X}$ .With every μ we associate the real-valued function on $\mathscr{A}$:

$$q_\mu(A) = \sup_{x \in X} \mu(T(x; A)). \tag{5.1.1}$$

DEFINITION 5.1.1

The function q_μ is called *cn.* (*with respect to* $\mathscr{A}$ *and* T).

Clearly q_μ is bounded by 1.
Suppose now that $\phi \in \mathscr{A}$, and:

(T_1) $T(x; A) = \varnothing$ iff $A = \varnothing$ for every $x \in X$;

(T_2) $T(x; A') \subset T(x; A'')$ and $T(x; A') \neq T(x; A'')$ for strict inclusion $A' \subset A''$;

(T_3) There exists an increasing sequence $(A_n)_{n \in N^*}$, $A_n \uparrow X$, such that $T(x; A_n) \uparrow X$ for at least one $x \in X$;

(T_4) $T(x; A_n) \uparrow T(x; A)$ for every $x \in X$ and for every increasing sequence $(A_n)_{n \in N^*}$, $A_n \uparrow A$.

If (T_1)–(T_4) hold, we have

THEOREM 5.1.2

q_μ has the following properties:

(q_1) $q_\mu(A) = 0$ if $A = \phi$;

(q_2) $q_\mu(A') \leq q_\mu(A'')$ for $A' \subset A''$;

(q_3) $q_\mu(X) = 1$;

(q_4) $\lim\limits_{A_n \uparrow A} q_\mu(A_n) = q_\mu(A)$ for every increasing sequence $(A_n)_{n \in N^*}$.

Proof

(q_1) and (q_2) are trivial.

(q_3) We have $1 \geq q_\mu(A_n) \geq \mu(T(x; A_n)) \uparrow \mu(X) = 1$ for the sequence referred to in (T_3); hence $q_\mu(X) = 1$.

(q_4) Let us write $q_\mu(A) < \mu(T(x_\varepsilon; A)) + \varepsilon$, and $q_\mu(A_n) \geq \mu(T(x_\varepsilon; A_n))$ for every $A_n \uparrow A$. Then by (T_4) we get $q_\mu(A) - q_\mu(A_n) \leq \varepsilon$. ◇

Assume further that we are interested in A's that can be characterized by a set of real parameters, then via q_μ we obtain a real-valued function of these parameters. There are clearly many such functions. If the set of parameters is determined, different A's will also lead to different functions with the same number of parameters.

More precisely, suppose that $\mathscr{A} = \{A(\boldsymbol{l}) : \boldsymbol{l} \in R^r\}$. It follows that T becomes a mapping from $X \times R^r$ into $\mathscr{X}$ and we put $T(x; \boldsymbol{l})$ for $T(x; A(\boldsymbol{l}))$. Next, we define by means of (5.1.1) the real-valued function Q_μ on R^r by setting

$$\bar{Q}_\mu(\boldsymbol{l}) = q_\mu(A(\boldsymbol{l})), \qquad \boldsymbol{l} \in R^r. \tag{5.1.2}$$

DEFINITION 5.1.3

The function $\bar{Q}_\mu$ is called a *cn.f.* (*with respect to* $\mathscr{A}$ *and* T).

To derive some properties of $\bar{Q}_\mu$ as defined by (5.1.2), we restrict ourselves to $A(\boldsymbol{l})$ that are compatible with (T_1)–(T_4), i.e.,

- (T_1) $T(x; \boldsymbol{l}) = \varnothing$ iff $\boldsymbol{l} \notin R^r_+$ for every $x \in X$;
- (T_2) $T(x; \boldsymbol{l}') \subset T(x; \boldsymbol{l}'')$ and $T(x; \boldsymbol{l}') \neq T(x; \boldsymbol{l}'')$ for $\boldsymbol{l}' < \boldsymbol{l}''$ for every $x \in X$;
- (T_3) there exists a sequence $(\boldsymbol{l}_n)_{n \in N^*}$, $\boldsymbol{l}_n \uparrow \infty$, such that $T(x; \boldsymbol{l}_n) \uparrow X$ for at least one $x \in X$;
- (T^2) $T(x; \boldsymbol{l}_n) \uparrow T(x; \boldsymbol{l})$ for every $x \in X$ and for every sequence $(\boldsymbol{l}_n)_{n \in N^*}$, $\boldsymbol{l}_n \uparrow \boldsymbol{l}$.

Clearly, we have

THEOREM 5.1.4

$\bar{Q}_\mu$ has the following properties:

- (Q_1) $\bar{Q}_\mu(\boldsymbol{l}) = 0$ for every $\boldsymbol{l} \notin R^r_+$;
- (Q_2) $\bar{Q}_\mu(\boldsymbol{l}') \leq \bar{Q}_\mu(\boldsymbol{l}'')$ for $\boldsymbol{l}' < \boldsymbol{l}''$;
- (Q_3) $\bar{Q}_\mu(\infty) = 1$;
- (Q_4) $\bar{Q}_\mu$ is left continuous.

We conclude that $\bar{Q}_\mu$ behaves almost like a d.f.; simple examples show however that $\bar{Q}_\mu$ is not always a d.f. since its rth difference may be negative (see p. 31). However, for $r = 1$, i.e., for $l \in R$, we get

THEOREM 5.1.5

$\bar{Q}_\mu$ is a (left continuous) d.f.

It should be noted that here we used a slightly different notation for the cn.f., this leads to left continuity.

5.1.2 As an illustration, let us take $(X, \mathscr{X}) = (R^r, \mathscr{B}^r)$. Consider the set $A^{\square}(\mathbf{l}) = \{\mathbf{z}: -\mathbf{l}/2 < \mathbf{z} < \mathbf{l}/2\}$; then we get the rectangular cn.f $Q_\mu^{\square -}(\mathbf{l}) = q_\mu(A^{\square}(\mathbf{l}))$.

In the next section we shall consider p.m.'s on metric spaces whose cn.f.'s depend on a single parameter.

Complement

Let us consider yet another cn.f. We start with the remark that the cn. of μ, as given by (5.1.1), can be written in the form

$$q_\mu(A) = \sup_{x \in X} \int_X \chi_{T(x;\, A)}(u)\, d\mu(u).$$

Denote now by $L_\mu^1(X)_+ = L^1(X, \mathscr{X}, \mu)_+$ the set of all real-valued non-negative measurable functions v on X which are μ-integrable, and denote by V a mapping from $\mathscr{A} \subset \mathscr{X}$ into the parts of $L_\mu^1(X)_+$. The function

$$q_\mu(A) = \sup_{v \in V(A)} \int_X v(u)\, d\mu(u), \quad A \in \mathscr{A}$$

is called *generalized cn. of* μ *(with respect to* $\mathscr{A}$ *and* V*)*. Clearly, if we take $V(A) = \{\chi_{T(x;\, A)}: x \in X\}$, then the generalized cn. reduces to that described by Definition 5.1.1. Further, the function $\bar{Q}_\mu(\mathbf{l}) = q_\mu(A(\mathbf{l}))$ *is called generalized cn.f. of* μ *(with respect to* $\mathscr{A}$ *and* V*)*. It is easily seen that from this generalized cn.f. we can obtain all cn.f.'s discussed in what preceeds [Hengartner and Theodorescu (1972a)].

Notes and Comments

The results are mainly to be found in Hengartner and Theodorescu (1972a).

5.2 CONCENTRATION FUNCTIONS IN METRIC SPACES

5.2.1 Let us assume in what follows that X is a metric space with the distance d, and let us take as $\mathscr{X}$ the family $\mathscr{B}_x$ of Borel sets of X. In order to get further results, let us focus our attention, e.g., on $\bar{Q}_\mu^\bigcirc$, i.e.,

$$\bar{Q}_\mu^\bigcirc(l) = \begin{cases} 0 & \text{if } \ l \le 0, \\ \sup\limits_{x \in X} \mu(S(x; l)) & \text{if } \ l > 0, \end{cases}$$

where $S(x; l)$ is the open sphere with center x and radius $l/2$.

THEOREM 5.2.1

We have

(1) $\bar{Q}_\mu^\bigcirc$ is a (left continuous) d.f.;

(2) $\bar{Q}_\mu^{\bigcirc +}(0) = \sup\limits_{x \in X} \mu(\{x\})$;

(3) $\bar{Q}_\mu^{\bigcirc +}(0) = 0$ iff μ is nonatomic;

(4) $\bar{Q}_\mu^\bigcirc = H_0^-$ iff $\mu = \delta_a$.

Moreover, if X is a normed vector space, then we have also

(5) $\bar{Q}_\mu^\bigcirc = \bar{Q}_{\mu_a}^\bigcirc = \bar{Q}_{\tilde{\mu}}^\bigcirc$, where $\mu_a(A) = \mu(A - a)$, and $\tilde{\mu}(A) = \mu(-A)$ for every $A \in \mathscr{B}_X$;

(6) $\bar{Q}^{\circ}_{\mu*\nu} \leq \min(\bar{Q}^{\circ}_{\mu}, \bar{Q}^{\circ}_{\nu})$;

(7) $\mu_{(n)}$ is nonatomic iff at least one of the μ_k is nonatomic.

(8) $\bar{Q}^{\circ}_{\mu*\nu}(l_1 + l_2) \geq \bar{Q}^{\circ}_{\mu}(l_1) \cdot \bar{Q}^{\circ}_{\nu}(l_2)$ for every $l_1, l_2 > 0$.

Proof

(1) follows directly from Theorem 5.1.4.

(2) (a) If $x_0 \in X$ such that $\mu(\{x_0\}) = a > 0$, then $\bar{Q}^{\circ}_{\mu}(l) \geq a$ for all $l > 0$, i.e., $\bar{Q}^{\circ +}(0) \geq \sup_{x \in X} \mu(\{x\})$.

(b) Assertion (2) is trivial if $\bar{Q}^{\circ +}_{\mu}(0) = 0$. Hence let $\bar{Q}^{\circ +}_{\mu}(0) = a > 0$. Then there exists a sequence $(x_k)_{k \in N^*}$ such that for any $\varepsilon > 0$ given, $\mu(S(x_k; 4^{-k})) > a - \varepsilon$. We shall show that $(x_k)_{k \in N^*}$ contains a convergent subsequence. Indeed, there is a subsequence $(k_n)_{n \in N^*}$ such that

$$S(x_{k_n}; 4^{-k_n}) \cap S(x_{k_{n+1}}; 4^{-k_{n+1}}) \neq \varnothing. \tag{5.2.1}$$

In fact, if not, we would have for infinitely many k_p

$$S(x_{k_p}; 4^{-k_p}) \cap \bigcup_{k_p < k} S(x_k; 4^{-k}) = \varnothing$$

and

$$\mu\left(\bigcup_{p \in N^*} S(x_{k_p}; 4^{-k_p})\right) = \sum_{p \in N^*} \mu(S(x_{k_p}; 4^{-k_p})) = \infty.$$

Consequently (5.2.1) implies $d(x_{k_m}, x_{k_n}) \leq 4^{-n_0+1}$ for $m, n \geq n_0$ and $S(x_{k_{n+p}}; 4^{-k_{n+p}}) \subset S(x_{k_n}; 4^{-n+1})$ for $p \in N^*$. Denote now $\bar{X}$ the completion of X and by $\bar{\mu}$ the induced p.m. defined on $\mathscr{B}_{\bar{X}}$ by $\bar{\mu}(A) = \mu(A \cap X)$, $A \in \mathscr{B}_{\bar{X}}$. Since $(x_{n_k})_{k \in N^*}$ is a Cauchy sequence, there is $x_0 \in \bar{X}$ such that

$$\lim_{k \to \infty} d(x_{n_k}, x_0) = 0,$$

and since for every $\delta > 0$ there is a natural number $n(\delta)$ such that $S(x_{n_k}; 4^{-n_k}) \subset S(x_0; \delta)$ if $n_k \geq n(\delta)$, we have for all $\varepsilon > 0$ and $\delta > 0$, $\bar{\mu}(S(x_0; \delta)) \geq a - \varepsilon$. Next, regularity of $\bar{\mu}$ implies $\bar{\mu}(\{x_0\}) \geq a$. But $\bar{\mu}(\{x_0\}) = \mu(\{x_0\} \cap X) \geq a$, so that it follows that $x_0 \in X$, and $\bar{Q}^{\circ +}_{\mu}(0) \leq \mu(\{x_0\})$.

(3) and (4) are straightforward consequences of (2), and (5) and (6) follow by the definition.

(7) If at least one μ_k is nonatomic, then we have $\bar{Q}^{\bigcirc +}_{\mu(n)}(0) \leq \bar{Q}^{\bigcirc +}_{\mu_k}(0) = 0$ and therefore by (2) it follows that $\mu_{(n)}$ is nonatomic. On the other hand, we have

$$\bar{Q}^{\bigcirc}_{\mu * \nu}(l) \geq \sup_{x \in X} \sup_{y \in X} \mu(S(x-y; l))\nu(\{y\})$$
$$= \bar{Q}^{\bigcirc}_{\mu}(l)\bar{Q}^{\bigcirc +}_{\nu}(0).$$

If $\mu_{(n)}$ is nonatomic, then

$$0 = \bar{Q}^{\bigcirc +}_{\mu(n)}(0) \geq \prod_{k=1}^{n} \bar{Q}^{\bigcirc +}_{\nu}(0),$$

so that, by (2), it follows that at least one μ_k is nonatomic.

(8) Let ξ and η be X-valued i.r.v.'s and denote respectively by μ and ν the p.m.'s induced on $\mathscr{B}_X$. For any $\varepsilon > 0$, there are $x, y \in X$ such that $P(\xi - x \in S(0; l_1)) > \bar{Q}^{\bigcirc}_{\mu}(l_1) - \varepsilon$ and $P(\eta - y \in S(0; l_2)) > \bar{Q}^{\bigcirc}_{\nu}(l_2) - \varepsilon$. Hence

$$(\bar{Q}^{\bigcirc}_{\mu}(l_1) - \varepsilon)(\bar{Q}^{\bigcirc}_{\nu}(l_2) - \varepsilon) \leq P(\{\xi - x \in S(0; l_1)\} \cap \{\eta - y \in S(0; l_2)\})$$
$$\leq P(\xi + \eta - (x+y) \in S(0; l_1 + l_2))$$
$$\leq \bar{Q}^{\bigcirc}_{\mu * \nu}(l_1 + l_2)$$

for any $\varepsilon > 0$, i.e., we obtain the desired property. ◇

Further, we have

THEOREM 5.2.2

Suppose X separable.† We have $\bar{Q}^{\bigcirc}_{\mu}(l) > 0$ for any $l > 0$.

Proof

For any $l > 0$, there is a countable set of spheres $S(x_k; l)$, $k \in N^*$, such that $X = \bigcup_{k \in N^*} S(x_k; l)$, and therefore for at least one $S(x_k; l)$ we have $\mu(Sx_k; l)) > 0$. ◇

† X is separable if it contains a countable dense subset; in other words, if each open cover of each subset of X has a countable subcover.

5.2.2 Let $\mathfrak{M} = \mathfrak{M}(X, \mathscr{B}_X)$ be the set of all p.m.'s on $\mathscr{B}_X$. For every $\mu, \nu \in \mathfrak{M}$, let us set

$$d^{\mathrm{P}}(\mu, \nu) = \inf A(\mu, \nu),$$

where

$$A(\mu, \nu) = \{\varepsilon > 0 : \nu(A) \leq \mu(A^\varepsilon) + \varepsilon,\ \mu(A) \leq \nu(A^\varepsilon) + \varepsilon \text{ for all } A \in \mathscr{B}_X\}.$$

and

$$A^\varepsilon = \{x : d(x, A) < \varepsilon\}.\dagger$$

Then $(\mathfrak{M}, d^{\mathrm{P}})$ is a metric space (Billingsley, 1968, p. 238), d^{P} being the Prohorov distance. Moreover, if X is separable, convergence in this metric is equivalent to the weak convergence [Dudley (1968, p. 1564)]. In particular, for $X = R^r$, d^{L} and d^{P} are equivalent.

Let us give now several results concerning convergence.

THEOREM 5.2.3

Let X be separable. Suppose that $\mu_n \in \mathfrak{M}$, $n \in N^*$, and

$$\mu_n \xrightarrow{\mathrm{w}} \mu \in \mathfrak{M} \qquad \text{as} \qquad n \to \infty.$$

Then $\bar{Q}_n^{\bigcirc} \xrightarrow{\mathrm{w}} \bar{Q}_\mu^{\bigcirc}$ as $n \to \infty$.

Proof

It suffices to show that $d^{\mathrm{L}}(\bar{Q}_\mu^{\bigcirc}, \bar{Q}_\nu^{\bigcirc}) \leq 2d^{\mathrm{P}}(\mu, \nu)$. In fact, we have for all $\varepsilon > d^{\mathrm{P}}(\mu, \nu)$, all $x \in X$, and all $l > 0$

$$\nu(S(x; l)) \leq \mu(S(x; l + 2\varepsilon)) + \varepsilon$$

and

$$\mu(S(x; l)) \leq \nu(S(x; l + 2\varepsilon)) + \varepsilon,$$

† It should be noted that we may replace A^ε by

$$A^{\varepsilon]} = \{x : d(x, A) \leq \varepsilon\}$$

in the definition of d^{P} without changing its value.

wherefrom it follows that $\bar{Q}_\nu^{\bigcirc}(l) \leq \bar{Q}_\mu^{\bigcirc}(l+2\varepsilon) + \varepsilon$ for all $l > 0$, and $\bar{Q}_\mu^{\bigcirc}(l) \leq \bar{Q}_\nu^{\bigcirc}(l+2\varepsilon) + \varepsilon$ for all $l > 0$. Hence $d^{\mathrm{L}}(\bar{Q}_\mu^{\bigcirc}, \bar{Q}_\nu^{\bigcirc}) \leq 2\varepsilon$, therefore $d^{\mathrm{L}}(\bar{Q}_\mu^{\bigcirc}, \bar{Q}_\nu^{\bigcirc}) \leq 2d^{\mathrm{P}}(\mu, \nu)$. ◇

We are now interested in the converse of Theorem 5.2.3. The same proof given for Theorem 3.3.7 holds for

THEOREM 5.2.4

Let X be a separable normed space. Then $\bar{Q}_n^{\bigcirc} \xrightarrow{\mathrm{w}} H_0^-$ as $n \to \infty$ iff μ_n is weakly essentially convergent to δ_0 as $n \to \infty$.

Further, we have

THEOREM 5.2.5

Let X be a separable Banach space. If $\sum_{n \in N^*} (1 - \bar{Q}_n^{\bigcirc +}(0)) < \infty$ [i.e., $\prod_{n \in N^*} \bar{Q}_n^{\bigcirc +}(0)$ converges], then $\mu_{(n)}$ is weakly essentially convergent.

Proof

Since $\sum_{n \in N^*} (1 - \bar{Q}_n^{\bigcirc +}(0)) < \infty$, there is a natural number $m_0(\varepsilon)$ for every $\varepsilon > 0$ such that

$$1 - \varepsilon \leq \prod_{k \geq m} (\bar{Q}_k^{\bigcirc +}(0)) < \bar{Q}_{(m,\infty)}^{\bigcirc +}(0)$$

and therefore $\bar{Q}_{(m,n)}^{\bigcirc +}(0) > 1 - \varepsilon$ for any $m, n \geq m_0(\varepsilon)$, and also $\lim_{m \to \infty} \bar{Q}_{(m,\infty)}^{\bigcirc} = H_0^-$. By Lemma 3.3.1 and Proposition 3.3.2 there is an integer-valued increasing function $\varphi(m) > m$, $m \in N^*$, such that $\lim_{m \to \infty} \bar{Q}_{(m)\,\varphi(m))}^{\bigcirc} = H_0^-$. Now the proof goes straightforward by modifying slightly the proof of Theorem 3.3.3. ◇

Notes and Comments

The results of this section are mainly due to Hengartner and Theodorescu; see also Diaz (1970) for Theorem 5.2.2.

5.3 CONCENTRATIONS IN HILBERT SPACES

5.3.1 Let X be a real separable Hilbert space, and let (x, y), $x, y \in X$ denote the inner product between x and y. Further, X endowed with the norm $\|x\| = (x, x)^{1/2}$ is a complete and separable normed vector space.

Let now μ be a p.m. on $\mathscr{B}_X$ and suppose that $\int_X \|x\| \, d\mu(x) < \infty$. Then $\int_X (x, y) \, d\mu(x)$ is well defined for every $y \in X$, and there is an $x_0 \in X$ such that $(x_0, y) = \int_X (x, y) \, d\mu(x)$ for every $y \in X$. The element x_0 is the *expectation* of μ, and we write $x_0 = E\mu = E\xi = \int_X x \, d\mu(x)$, where ξ is the corresponding X-valued r.v. Clearly, μ_{x_0} has expectation 0.

Further, suppose that $\int_X \|x\|^2 d\mu(x) < \infty$. Then the covariance operator S of μ is the Hermitian operator uniquely defined by the quadratic form

$$(Sy, y) = \int_X (x, y)^2 \, d\mu(x).$$

Moreover, the trace of S is defined by

$$\operatorname{tr} S = \sum_{n \in N^*} (Se_n, e_n),$$

where $(e_n)_{n \in N^*}$ is a complete orthonormal basis for X.

5.3.2 We start by proving

THEOREM 5.3.1

Suppose that $\int_X \|x\|^2 \, d\mu(x) < \infty$.

Then

$$1 - \operatorname{tr} S/4l^2 \leq \bar{Q}_\mu^{\bigcirc}(l)$$

for every $l > 0$.

Proof

In fact we can write

$$\operatorname{tr} S = \sum_{n \in N^*} (Se_n, e_n) - \sum_{n \in N^*} \int_X (x, e_n)^2 \, d\mu(x) = \int_X \|x\|^2 \, d\mu(x)$$

$$\geq \int_{X \cap S^c(0;l)} \|x\|^2 \, d\mu(x) \geq l^2(1 - \bar{Q}_\mu^{\bigcirc}(l))/4$$

for every $l > 0$. ◇

Next, we have the following result:

THEOREM 5.3.2

If $\mu_1 = \cdots = \mu_n = \mu$, we have

$$\bar{Q}_{\mu(n)}(l) \leq A(\mu; l)n^{-1/2}$$

where $A(\mu; l)$ is a finite positive constant independent of n but depending on μ and on l iff $\mu \neq \delta_{x_0}$.

Proof

If $\mu = \delta_{x_0}$, then $\bar{Q}_{\mu(n)}^{\bigcirc} = H_0^-$.

Let now $\mu \neq \delta_{x_0}$. Then there exists $y \neq x_0$, $y \in S_\mu$, S_μ being the support of μ. Since $\bar{Q}_\mu^{\bigcirc}$ is invariant under shift, we may assume $x_0 = 0$. Further, let $(e_n)_{n \in N^*}$ be a complete orthonormal basis for X, where $e_1 = y/\|y\|$. Define $\mu_1((-\infty, a)) = \mu(\{x : -\infty < (x, e_1) < a\})$, $a \in R$. Clearly, μ_1 is a p.m. on $\mathscr{B}$, and $\mu_1 \neq \delta_a$. We have

$$\bar{Q}_{\mu * \nu}^{\bigcirc +}(l) \leq \sup_{\lambda \in R} \mu * \nu(\{x : \lambda - l/2 \leq (x, e_1) \leq \lambda + l/2\})$$

$$= Q_{(\mu * \nu)_1}(l).$$

But

$$(\mu * \nu)_1([\lambda - l/2, \lambda + l/2])$$

$$= \int \mu(\{x : \lambda - (t, e_1) - l/2 \le (x, e_1) \le \lambda - (t, e_1) + l/2\})\, d\nu(t)$$

$$= \int \mu_1([\lambda - (t, e_1) - l/2, \lambda - (t, e_1) + l/2])\, d\nu(t)$$

$$= \int_R \mu_1([\lambda - t_1 - l/2, \lambda + l/2]).$$

Therefore, we get, using (2.2.35), $\bar{Q}^{\bigcirc +}_{\mu(n)}(l) \le Q_{(\mu_{(n)})_1}(l) = Q_{(\mu_1)(n)}(l) \le A(\mu; l)n^{-1/2}$. $\diamond$

We get immediately the following consequence.

THEOREM 5.3.3

We have $\bar{Q}^{\bigcirc}_{\mu * \nu} = \bar{Q}^{\bigcirc}_{\mu}$ iff $\nu = \delta_a$.

5.3.3 Let S be the Hermitian operator defined by $(Se_j, e_k) = 0, j \ne k$, and $(Se_j, e_j) = 2^{-j}$. Then there is a p.m. λ on $\mathscr{B}_X$ such that the c.f. of λ, ψ_λ, can be written under the term

$$\exp\{-\tfrac{1}{2}(St, t)\} = \psi_\lambda(t) = \int_X e^{i(t,x)}\, d\lambda(x).$$

The natural generalization of Definition 4.1.1 is

DEFINITION 5.3.4

The real number

$$q_\mu = \int_X |\psi_\mu(t)|^2\, d\lambda(t),$$

where ψ_λ is the c.f. of the p.m. μ, is called the *convergence cn. of* μ *(with respect to* λ*)*.

Let us give some properties of q_μ.

THEOREM 5.3.5

We have

(1) $q_\mu = q_{\mu_a} = q_{\tilde{\mu}}$;
(2) $0 < \exp(-2l^2)(\bar{Q}_\mu^{\bigcirc}(l))^2 \le q_\mu \le 1$ for all $l > 0$;
(3) $q_\mu = 1$ iff $\mu = \delta_{x_0}$;
(4) $q_{\mu * \nu} \le \min(q_\mu, q_\nu)$;
(5) $q_{\mu_n} \to q_\mu$ as $n \to \infty$ if μ_n is weakly essentially convergent to μ as $n \to \infty$.

Proof

(1), (4), and (5) are straightforward consequences of the definition.
(2) We have

$$1 \ge \int_{X \times X} \exp\{-\tfrac{1}{2}(S(x-y), x-y)\, d\mu(x)\, d\mu(y) = q_\mu$$
$$\ge \int_{S(0;\, l) \times S(0;\, l)} \exp(-l^2/2)\, d\mu(x)\, d\mu(y) = \exp(-l^2/2)(\mu(S(0; l)))^2$$

for all $l > 0$. But since $q_{\mu_a} = q_\mu$, we have $q_\mu \ge \exp(-l^2/2)(\bar{Q}_\mu^{\bigcirc}(l) - \varepsilon)^2$ for every $\varepsilon > 0$, therefore we get (2).

(3) If $\mu = \delta_{x_0}$, then $q_\mu = 1$ by definition. On the other hand, $q_\mu = 1$ iff $q_{\tilde{\mu}} = 1$ since $|\psi_\mu| = |\psi_{\tilde{\mu}}| = 1$ λ – a.s. Let $\tilde{\mu} \ne \delta_0$, and denote by $\tilde{\mu}_k$, $k \in N^*$, the p.m.'s generated by $\tilde{\mu}_k((-\infty, a)) = \tilde{\mu}(\{x : -\infty < (x, l_k) < a\})$. Then there is at least one k with $\tilde{\mu}_k \ne \delta_0$. Therefore we have an $l > 0$ such that

$$q_\mu = \int_X \exp(-\tfrac{1}{2}(St, t)\, d\tilde{\mu}(t) \le \int_X \exp(-2^{-k}t_k^2)\, d\tilde{\mu}(t)$$
$$\le \exp(-2^{-k}l^2) \int_{|(y,\, l_k)| \ge l} d\tilde{\mu}(y) + \int_{|(y,\, l_k)| < l} d\tilde{\mu}(y) < 1,$$

and we are led to a contradiction. ◇

However, it is not true that $q_{\mu_n} \to 1$ as $n \to \infty$ implies μ_n is essentially weakly convergent to δ_0 as $n \to \infty$. In fact, take $\mu_n = \frac{1}{2}\delta_{e_n} + \frac{1}{2}\delta_{-e_n}$, $n \in N^*$.

Complement

In order to carry out proofs of limit theorems in a Hilbert space it is necessary to get some generalization of Kolmogorov inequality [see, e.g., Loève (1963, p. 235)]. This is done in terms of cn.f.'s. First, (3.3.4) holds true. Suppose now that $1 \leq k \leq n$, are symmetric i.r.v.'s such that $\|\xi_k\| \leq c$, $1 \leq k \leq n$. Then

$$E\|\xi^2_{(n)}\| \leq (16l^2 + (c + 4l)^2)/(2\bar{Q}^{\bigcirc +}_{(n)}(l) - 1)$$

for any l such that $\bar{Q}^{\bigcirc}_{(n)}(l) > \frac{1}{2}$ [Parthasarathy (1967, p. 168–170)].

Notes and Comments

The results of this section are due to Hengartner and Theodorescu.

REFERENCES

Bahr, B. von, and Esséen, C. G.

(1965) Inequalities for the rth absolute moment of a sum of random variables, $1 \leq r \leq 2$. *Ann. Math. Statist.* **36**, 299–303.

Billingsley, P.

(1968) Convergence of Probability Measures." Wiley, New York.

Chung, K. L., and Erdös, P.

(1951) Probability limit theorems assuming only the first moment I. *Mem. Amer. Math. Soc.* **6**, 1–19.

Csiszár, I.

(1962) Informationstheoretische Konvergenzbegriffe im Raum der Wahrscheinlichkeitsverteilungen. *Magyar. Tud. Akad. Mat. Kut. Int. Közl.* **7**, 137–158.

(1963) Eine informationstheoretische Ungleichung und ihre Anwendung auf den Beweis der Ergodizität von Markoffschen Ketten. *Publ. Math. Inst. Hung. Acad. Sci.* **A8**, 85–108.

Dachuna-Castelle, D., Revuz, D., and Schreiber, M.

(1970) "Recueil de problemes de calcul des probabilités," Masson, Paris.

Diaz, M. M.

(1970) Algunas cwestiones solere la función de concentración de una probabilidad *Trabajos Estadist.*, **64**, 139–160.

Doeblin, W.

(1939) Sur les sommes d'un grand nombre de variables aléatoires indépendantes. *Bull. Sci. Math.* **63**, 23–64.

Doeblin, W., and Lévy, P.

(1936) Sur les sommes de variables aléatoires indépendantes à dispersion bornées inférieurement. *C. R. Acad. Sci. Paris.* **202**, 2027–2029.

Dudley, R. M.

(1968) Distances of probability measures and random variables. *Ann. Math. Statist.* **39**, 1563–1572.

Dunnage, J. E. A.

(1968) The number of real zeros of a class of random algebraic polynomials. *Proc. London Math. Soc.* **18**, 439–460.

(1971) Inequalities for the concentration functions of sums of independent random variables. *Proc. London Math. Soc.* **23**, 489–515.

Erdös, P.

(1945) On a lemma of Littlewood and Offord. *Bull. Amer. Math. Soc.* **51**, 898–902.

Esséen, C. G.

(1945) Fourier analysis of distribution functions. *Acta Math* **77**, 1–125.

(1966) On the Kolomogorov–Rogozin inequality for the concentration function. *Z. Wahrscheinlichkeitstheorie Verw. und* Gebiete **5**, 210–216.

(1968) On the concentration function of a sum of independent random variables. *Z. Wahrscheinlichkeitstheorie Verw. und Gebiete* **9**, 290–308.

Gnedenko, B. V., and Kolmogorov, A. N.

(1954) "Limit Distributions for Sums of Independent Random Variables." Addison-Wesley, New York.

Grenander, U.

(1963) "Probabilities on Algebraic Structures." Wiley, New York.

Hengartner, W., and Theodorescu, R.

(1977a) Concentration functions I. Symposia Mathematica 9, Rome, 15–18 March 1971, 79–106, Academic Press, New York.

(1972b) Concentration functions II. *Atti. Accad. Naz. Lincei, Rend. Cl. Sci. Fis. Mat. Natur.*, Ser. **8**, **52**, 240–243.

Heyde, C. C.

(1966) Some results on small-deviation probability convergence rates for sums of independent random variables. *Canad. J. Math.* **18**, 656–665.

Hille, E., and Phillips, R. S.

(1957) "Functional analysis and semi-groups." Amer. Math. Soc., Providence, Rhode Island.

Hinčin (Khintchine), A. Ja.

(1937) Sur l'arithmétique des lois de distribution. *Bull. Univ. Moscow* **1**, 6–17.

Ito, K.

(1942) On stochastic processes. I. *Jap. J. Math.* **18**, 261–301.

(1960) Stochastic processes; Part I. *Izd. Innostrannoi lit.*, Moscow (in Russian).

Jessen, B., and Wintner, A.

(1935) Distribution functions and the Riemann zeta function. *Trans. Amer. Math. Soc.* **38**, 48–88.

Kesten, H.

(1969) A sharper form of the Doeblin–Lévy–Kolmogorov–Rogozin inequality for concentration functions. *Math. Scand.* **25**, 133–144.

(1972) Sums of independent random variables–without moment conditions. *Ann. Math. Statist.* **43**, 701–732.

Kolmogorov, A. N.

(1956) Two uniform limit theorems for sums of independent random variables. *Theor. Prob. Appl.* **1**, 384–394.

(1958) Sur les proprietes des fonctions de concentration de M. P. Lévy. *Ann. Inst. H. Poincaré, Sect. B.* **16**, 27–34.

(1963) On the approximation of distributions of sums of independent summands by infinitely divisible distributions. *Sankhya Ser. A.* **25**, 159–174.

Kullback, S., and Leibler, R. A.
(1951) On information and sufficiency. *Ann. Math. Statist.* **22**, 79–86.

Lecam, L.
(1963) A note on the distribution of sums of independent random variables. *Proc. Nat. Acad. Sci. U.S.A.* **50**, 601–603.
(1965a) On the distribution of sums of independent random, *in* "Bernoulli (1713), Bayes (1763), Laplace (1813)." J. Neyman and L. LeCam, (eds.) 179–202. Springer-Verlag, New York.
(1965b) A remark on the central limit theorem. *Proc. Nat. Acad. Sci. U.S.A.* **54**, 354–359.

Lévy, P.
(1937) "Théorie de l'addition des variables aléatoines," Gauthier-Villars, Paris.
(1954) "Théorie de l'addition des variables aléatoines," 2nd Ed., Gauthier-Villars, Paris.

Linnik, Ju. V.
(1962) Décompositions des lois de probabilités. Gauthier-Villars, Paris.

Littlewood, J. E., and Offord, A. C.
(1943) On the number of real roots of a random algebraic equation III. *Mat. Sb.* **12**, 277–286.

Loève, M.
(1963) "Probability Theory," 3rd Ed. Van Nostrand, New York.

Lukacs, E.
(1970) "Characteristic Functions," 2nd Ed. Griffin, London.

Offord, A. C.
(1945) An inequality for sums of independent random variables. *Proc. London Math. Soc.* (2) **48**, 467–477.

Onicescu, O.
(1966) Énergie informationnelle. *C. R. Acad. Sci. Paris Sér. A* **263**, A841–A842.

Parthasarathy, K. R.
(1967) "Probability Measures on Metric Spaces." Academic Press, New York.

Pérez, A.
(1957) Notions généralisées d'incertitude, d'entropie et d'information du point de vue de la théorie des martingales. "Transactions of the First Prague Conference on Information Theory, Statistical Decision Functions, Random Processes, Prague 1956," pp. 183–208. Publ. House Czech. Acad. Sci., Prague.
(1967a) Information-theoretic risk estimates in statistical decision. *Kybernetika (Prague)* **3**, 1–21.
(1967b) Sur l'énergie informationnelle de M. Octav Onicescu. *Rev. Roumaine Math. Pures Appl.* **12**, 1314–1347.

Petrov, V. V.
(1970) On an estimate of the concentration function of a sum of independent random variables. *Theor. Prob. Appl.* **15**, 701–703.

Pinsker, M. S.
(1960) Information and informational stability of random variables and of random processes. *Izd. Akad. Nauk SSSR*, Moscow (In Russian).

Prohorov, Yu. V.
(1956) Convergence of random processes and limit theorems in probability theory. *Theor. Prob. Appl.* **1**, 157–214.

Prohorov, Yu. V., and Rozanov, Yu. A.
(1969) "Probability theory." Springer-Verlag, Berlin and New York.

Raoult, J. P.
(1969) Contiguité. Décomposition de Lebesgue des suites de fonctions σ-additives. Ph. D. Thesis, University of Paris, Paris.
(1970) Décomposition des suites de mesures. *Bull. Sci. Math.* **94**, 209–229.

Rényi, A.
(1961) On measures of entropy and information. "Proceedings of the Fourth Berkeley Symposium on Mathematical Statistics and Probability," Vol. I, pp. 547–561. Univ. California Press, Berkeley, 1961.

Rogozin, B. A.
(1961a) An estimate for concentration functions. *Theor. Prob. Appl.* **6**, 94–96.
(1961b) On the increase of dispersion of sums of independent random variables. *Theor. Prob. Appl.* **6**, 97–99.

Rosén, B.
(1961) On the asymptotic distribution of sums of independent random variables. *Ark. Mat.* **4**, 323–332.

Sâmboan, G., and Theodorescu, R.
(1968) On the notion of concentration I. *Elektron. Informationsverarbeit. Kybernetik* **4**, 235–255. (Note: No further parts of this paper will appear.)

Sazonov, V. V.
(1966) On multidimensional concentration functions. *Theor. Prob. Appl.* **11**, 603–609.

Sebast' yanov, B. A.
(1963) On multidimensional concentrations. *Theor. Prob. Appl.* **8**, 116.

Tranter, C. J.
(1968) "Bessel Functions with Some Physical Applications. Hart, New York.

Tucker, H. G.
(1963) Quasi-convergent series of independent random variables. *Amer. Math. Monthly* **70**, 718–722.

AUTHOR INDEX

Numbers in italics show the page on which the complete reference is listed.

SUBJECT INDEX

The contents should also be consulted for subject matter.